I0010855

Practical Data Analysis

Second Edition

A practical guide to obtaining, transforming, exploring, and analyzing data using Python, MongoDB, and Apache Spark

Hector Cuesta
Dr. Sampath Kumar

BIRMINGHAM - MUMBAI

Practical Data Analysis

Second Edition

Copyright © 2016 Packt Publishing

All rights reserved. No part of this book may be reproduced, stored in a retrieval system, or transmitted in any form or by any means, without the prior written permission of the publisher, except in the case of brief quotations embedded in critical articles or reviews.

Every effort has been made in the preparation of this book to ensure the accuracy of the information presented. However, the information contained in this book is sold without warranty, either express or implied. Neither the authors, nor Packt Publishing, and its dealers and distributors will be held liable for any damages caused or alleged to be caused directly or indirectly by this book.

Packt Publishing has endeavored to provide trademark information about all of the companies and products mentioned in this book by the appropriate use of capitals. However, Packt Publishing cannot guarantee the accuracy of this information.

First published: October 2013

Second published: September 2016

Production reference: 1260916

Published by Packt Publishing Ltd.

Livery Place

35 Livery Street

Birmingham B3 2PB, UK.

ISBN 978-1-78528-971-2

www.packtpub.com

Credits

Authors

Hector Cuesta
Dr. Sampath Kumar

Reviewers

Chandana N. Athauda
Mark Kerzner

Commissioning Editor

Amarabha Banarjee

Acquisition Editor

Denim Pinto

Content Development Editor

Divij Kotian

Technical Editor

Rutuja Vaze

Copy Editor

Safis Editing

Project Coordinator

Ritika Manoj

Proofreader

Safis Editing

Indexer

Tejal Daruwale Soni

Production Coordinator

Melwyn Dsa

Cover Work

Melwyn Dsa

About the Authors

Hector Cuesta is founder and Chief Data Scientist at Dataxios, a machine intelligence research company. Holds a BA in Informatics and a M.Sc. in Computer Science. He provides consulting services for data-driven product design with experience in a variety of industries including financial services, retail, fintech, e-learning and Human Resources. He is an enthusiast of Robotics in his spare time.

You can follow him on Twitter at `https://twitter.com/hmCuesta`.

I would like to dedicate this book to my wife Yolanda, and to my wonderful children Damian and Isaac for all the joy they bring into my life. To my parents Elena and Miguel for their constant support and love.

Dr. Sampath Kumar works as an assistant professor and head of Department of Applied Statistics at Telangana University. He has completed M.Sc., M.Phl., and Ph. D. in statistics. He has five years of teaching experience for PG course. He has more than four years of experience in the corporate sector. His expertise is in statistical data analysis using SPSS, SAS, R, Minitab, MATLAB, and so on. He is an advanced programmer in SAS and matlab software. He has teaching experience in different, applied and pure statistics subjects such as forecasting models, applied regression analysis, multivariate data analysis, operations research, and so on for M.Sc. students. He is currently supervising Ph.D. scholars.

About the Reviewers

Chandana N. Athauda is currently employed at BAG (Brunei Accenture Group) Networks—Brunei and he serves as a technical consultant. He mainly focuses on Business Intelligence, Big Data and Data Visualization tools and technologies.

He has been working professionally in the IT industry for more than 15 years (Ex-Microsoft Most Valuable Professional (MVP) and Microsoft Ranger for TFS). His roles in the IT industry have spanned the entire spectrum from programmer to technical consultant. Technology has always been a passion for him.

If you would like to talk to Chandana about this book, feel free to write to him at info `@inzeek.net` or by giving him a tweet `@inzeek`.

Mark Kerzner is a Big Data architect and trainer. Mark is a founder and principal at Elephant Scale, offering Big Data training and consulting. Mark has written *HBase Design Patterns* for Packt.

I would like to acknowledge my co-founder Sujee Maniyam and his colleague Tim Fox, as well as all the students and teachers. Last but not least, thanks to my multi-talented family.

www.PacktPub.com

For support files and downloads related to your book, please visit www.PacktPub.com.

eBooks, discount offers, and more

Did you know that Packt offers eBook versions of every book published, with PDF and ePub files available? You can upgrade to the eBook version at www.PacktPub.com and as a print book customer, you are entitled to a discount on the eBook copy. Get in touch with us at customercare@packtpub.com for more details.

At www.PacktPub.com, you can also read a collection of free technical articles, sign up for a range of free newsletters and receive exclusive discounts and offers on Packt books and eBooks.

https://www2.packtpub.com/books/subscription/packtlib

Do you need instant solutions to your IT questions? PacktLib is Packt's online digital book library. Here, you can search, access, and read Packt's entire library of books.

Why subscribe?

- Fully searchable across every book published by Packt
- Copy and paste, print, and bookmark content
- On demand and accessible via a web browser

Free access for Packt account holders

Get notified! Find out when new books are published by following @PacktEnterprise on Twitter or the Packt Enterprise Facebook page.

Table of Contents

Preface

Practical Data Analysis provides a series of practical projects in order to turn data into insight. It covers a wide range of data analysis tools and algorithms for classification, clustering, visualization, simulation and forecasting. The goal of this book is to help you to understand your data to find patterns, trends, relationships and insight.

This book contains practical projects that take advantage of the MongoDB, D3.js, Python language and its ecosystem to present the concepts using code snippets and detailed descriptions.

What this book covers

Chapter 1, *Getting Started,* In this chapter, we discuss the principles of data analysis and the data analysis process.

Chapter 2, *Preprocessing Data,* explains how to scrub and prepare your data for the analysis, also introduces the use of OpenRefine which is a Data Cleansing tool.

Chapter 3, *Getting to Grips with Visualization,* shows how to visualize different kinds of data using D3.js which is a JavaScript Visualization Framework.

Chapter 4, *Text Classification,* introduces the binary classification using a Naïve Bayes Algorithm to classify spam.

Chapter 5, *Similarity-Based Image Retrieval,* presents a project to find the Similarity between images using a dynamic time warping approach.

Chapter 6, *Simulation of Stock Prices,* explains how to simulate a Stock Price using Random Walk algorithm, visualized with a D3.js animation.

Chapter 7, *Predicting Gold Prices,* introduces how Kernel Ridge Regression works, and how to use it to predict the gold price using time series.

Chapter 8, *Working with Support Vector Machines,* describes how to use Support Vector Machines as a classification method.

Chapter 9, *Modeling Infectious Diseases with Cellular Automata,* introduces the basic concepts of Computational Epidemiology simulation and explains how to implement a cellular automaton to simulate an epidemic outbreak using D3.js and JavaScript.

Chapter 10, *Working with Social Graphs*, explains how to obtain and visualize your social media graph from Facebook using Gephi.

Chapter 11, *Working with Twitter Data*, explains how to use the Twitter API to retrieve data from twitter. We also see how to improve the text classification to perform a sentiment analysis using the Naïve Bayes Algorithm implemented in the Natural Language Toolkit (NLTK).

Chapter 12, *Data Processing and Aggregation with MongoDB*, introduces the basic operations in MongoDB as well as methods for grouping, filtering, and aggregation.

Chapter 13, *Working with MapReduce*, illustrates how to use the MapReduce programming model implemented in MongoDB.

Chapter 14, *Online Data Analysis with* Jupyter *and Wakari*, explains how to use the Wakari platform and introduces the basic use of Pandas and PIL with IPython.

Chapter 15, *Understanding Data Processing using Apache Spark*, explains how to use distributed file system along with Cloudera VM and how to get started with a data environment. Finally, we describe the main features of Apache Spark with a practical example.

What you need for this book

The basic requirements for this book are as follows:

- Python
- OpenRefine
- D3.js
- Mlpy
- Natural Language Toolkit (NLTK)
- Gephi
- MongoDB

Who this book is for

This book is for Software Developers, Analyst and Computer Scientists who want to implement data analysis and visualization in a practical way. The book is also intended to provide a self-contained set of practical projects in order to get insight from different kinds of data like, time series, numerical, multidimensional, social media graphs and texts.

You are not required to have previous knowledge about data analysis, but some basic knowledge about statistics and a general understanding of Python programming is assumed.

Conventions

In this book, you will find a number of text styles that distinguish between different kinds of information. Here are some examples of these styles and an explanation of their meaning. Code words in text, database table names, folder names, filenames, file extensions, pathnames, dummy URLs, user input, and Twitter handles are shown as follows: "For this example, we will use the `BeautifulSoup` library version 4."

A block of code is set as follows:

```
from bs4 import BeautifulSoup
import urllib.request
from time import sleep
from datetime import datetime
```

Any command-line input or output is written as follows:

```
>>> readers@packt.com
>>> readers
>>> packt.com
```

New terms and important words are shown in bold. Words that you see on the screen, for example, in menus or dialog boxes, appear in the text like this: "Now, just click on the **OK** button to apply the transformation."

 Warnings or important notes appear in a box like this.

Tips and tricks appear like this.

Reader feedback

Feedback from our readers is always welcome. Let us know what you think about this book—what you liked or disliked. Reader feedback is important for us as it helps us develop titles that you will really get the most out of.

To send us general feedback, simply e-mail `feedback@packtpub.com`, and mention the book's title in the subject of your message.

If there is a topic that you have expertise in and you are interested in either writing or contributing to a book, see our author guide at `www.packtpub.com/authors`.

Customer support

Now that you are the proud owner of a Packt book, we have a number of things to help you to get the most from your purchase.

Downloading the example code

You can download the example code files for this book from your account at `http://www.packtpub.com`. If you purchased this book elsewhere, you can visit `http://www.packtpub.com/support` and register to have the files e-mailed directly to you.

You can download the code files by following these steps:

1. Log in or register to our website using your e-mail address and password.
2. Hover the mouse pointer on the **SUPPORT** tab at the top.
3. Click on **Code Downloads & Errata**.
4. Enter the name of the book in the **Search** box.
5. Select the book for which you're looking to download the code files.
6. Choose from the drop-down menu where you purchased this book from.
7. Click on **Code Download**.

You can also download the code files by clicking on the **Code Files** button on the book's webpage at the Packt Publishing website. This page can be accessed by entering the book's name in the Search box. Please note that you need to be logged in to your Packt account.

Once the file is downloaded, please make sure that you unzip or extract the folder using the latest version of:

- WinRAR / 7-Zip for Windows
- Zipeg / iZip / UnRarX for Mac
- 7-Zip / PeaZip for Linux

The code bundle for the book is also hosted on GitHub at `https://github.com/PacktPubl ishing/Practical-Data-Analysis-Second-Edition`. We also have other code bundles from our rich catalog of books and videos available at `https://github.com/PacktPublish ing/`. Check them out!

Downloading the color images of this book

We also provide you with a PDF file that has color images of the screenshots/diagrams used in this book. The color images will help you better understand the changes in the output. You can download this file from `https://www.packtpub.com/sites/default/files/down loads/B4227_PracticalDataAnalysisSecondEdition_ColorImages.pdf`.

Errata

Although we have taken every care to ensure the accuracy of our content, mistakes do happen. If you find a mistake in one of our books—maybe a mistake in the text or the code—we would be grateful if you could report this to us. By doing so, you can save other readers from frustration and help us improve subsequent versions of this book. If you find any errata, please report them by visiting `http://www.packtpub.com/submit-errata`, selecting your book, clicking on the Errata Submission Form link, and entering the details of your errata. Once your errata are verified, your submission will be accepted and the errata will be uploaded to our website or added to any list of existing errata under the Errata section of that title.

To view the previously submitted errata, go to `https://www.packtpub.com/books/conten t/support` and enter the name of the book in the search field. The required information will appear under the Errata section.

Piracy

Piracy of copyrighted material on the Internet is an ongoing problem across all media. At Packt, we take the protection of our copyright and licenses very seriously. If you come across any illegal copies of our works in any form on the Internet, please provide us with the location address or website name immediately so that we can pursue a remedy.

Please contact us at `copyright@packtpub.com` with a link to the suspected pirated material.

We appreciate your help in protecting our authors and our ability to bring you valuable content.

Questions

If you have a problem with any aspect of this book, you can contact us at `questions@packtpub.com`, and we will do our best to address the problem.

1
Getting Started

Data analysis is the process in which raw data is ordered and organized to be used in methods that help to evaluate and explain the past and predict the future. Data analysis is not about the numbers, it is about making/asking questions, developing explanations, and testing hypotheses based on logical and analytical methods. Data analysis is a multidisciplinary field that combines computer science, artificial intelligence, machine learning, statistics, mathematics, and business domain, as shown in the following figure:

All of these skills are important for gaining a good understanding of the problem and its optimal solutions, so let's define those fields.

Computer science

Computer science creates the tools for data analysis. The vast amount of data generated has made computational analysis critical and has increased the demand for skills like programming, database administration, network administration, and high-performance computing. Some programming experience in Python (or any high-level programming language) is needed to follow the chapters in this book.

Artificial intelligence

According to Stuart Russell and Peter Norvig:

> *"Artificial intelligence has to do with smart programs, so let's get on and write some".*

In other words, **Artificial intelligence** (**AI**) studies the algorithms that can simulate an intelligent behavior. In data analysis we use AI to perform those activities that require intelligence, like **inference**, **similarity search**, or **unsupervised classification**. Fields like deep learning rely on artificial intelligence algorithms; some of its current uses are chatbots, recommendation engines, image classification, and so on.

Machine learning

Machine learning (**ML**) is the study of computer algorithms to learn how to react in a certain situation or recognize patterns. According to Arthur Samuel (1959):

> *"Machine Learning is a field of study that gives computers the ability to learn without being explicitly programmed".*

ML has a large amount of algorithms generally split into three groups depending how the algorithms are training. They are as follows:

- Supervised learning
- Unsupervised learning
- Reinforcement learning

The relevant number of algorithms is used throughout the book and they are combined with practical examples, leading the reader through the process from the initial data problem to its programming solution.

Statistics

In January 2009, Google's Chief Economist Hal Varian said:

> *"I keep saying the sexy job in the next ten years will be statisticians. People think I'm joking, but who would've guessed that computer engineers would've been the sexy job of the 1990s?"*

Statistics is the development and application of methods to collect, analyze, and interpret data. Data analysis encompasses a variety of statistical techniques such as **simulation, Bayesian methods, forecasting, regression, time-series analysis**, and **clustering**.

Mathematics

Data analysis makes use of a lot of mathematical techniques like **linear algebra** (vector and matrix, factorization, eigenvalue), numerical methods, and conditional probability, in algorithms. In this book, all the chapters are self-contained and include the necessary math involved.

Knowledge domain

One of the most important activities in data analysis is asking questions, and a good understanding of the knowledge domain can give you the expertise and intuition needed to ask good questions. Data analysis is used in almost every domain, including finance, administration, business, social media, government, and science.

Data, information, and knowledge

Data is facts of the world. Data represents a fact or statement of an event without relation to other things. Data comes in many forms, such as web pages, sensors, devices, audio, video, networks, log files, social media, transactional applications, and much more. Most of these data are generated in real time and on a very large-scale. Although it is generally alphanumeric (text, numbers, and symbols), it can consist of images or sound. Data consists of raw facts and figures. It does not have any meaning until it is processed. For example, financial transactions, age, temperature, and the number of steps from my house to my office are simply numbers. The information appears when we work with those numbers and we can find value and meaning.

Information can be considered as an aggregation of data. Information has usually got some meaning and purpose. The information can help us to make decisions easier. After processing the data, we can get the information within a context in order to give proper meaning. In computer jargon, a relational database makes information from the data stored within it.

Knowledge is information with meaning. Knowledge happens only when human experience and insight is applied to data and information. We can talk about knowledge when the data and the information turn into a set of rules to assist the decisions. In fact, we can't store knowledge because it implies the theoretical or practical understanding of a subject. The ultimate purpose of knowledge is for value creation.

Inter-relationship between data, information, and knowledge

We can observe that the relationship between data, information, and knowledge looks like cyclical behavior. The following diagram demonstrates the relationship between them. This diagram also explains the transformation of data into information and vice versa, similarly information and knowledge. If we apply valuable information based on context and purpose, it reflects knowledge. At the same time, the processed and analyzed data will give the information. When looking at the transformation of data to information and information to knowledge, we should concentrate on the context, purpose, and relevance of the task.

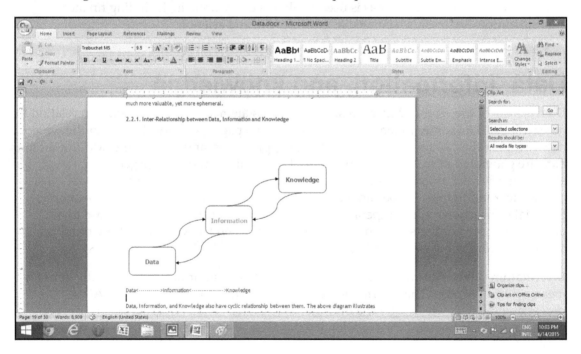

Now I would like to discuss these relationships with a real-life example:

Our students conducted a survey for their project with the purpose of collecting data related to customer satisfaction of a product and to see the conclusion of reducing the price of that product. As it was a real project, our students got to make the final decision to satisfy the customers. Data collected by the survey was processed and a final report was prepared. Based on the project report, the manufacturer of that product has since reduced the cost. Let's take a look at the following:

- **Data**: Facts from the survey.
  - For example: Number of customers purchased the product, satisfaction levels, competitor information, and so on.
- **Information**: Project report.
  - For example: Satisfaction level related to price based on the competitor product.
- **Knowledge**: The manufacturer learned what to do for customer satisfaction and increase product sales.
  - For example: The manufacturing cost of the product, transportation cost, quality of the product, and so on.

Finally, we can say that the `data-information-knowledge` hierarchy seemed like a great idea. However, by using predictive analytics we can simulate an intelligent behavior and provide a good approximation. In the following image is an example of how to turn data into knowledge:

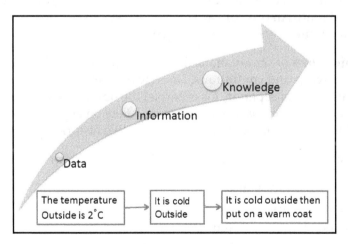

The nature of data

Data is the plural of datum, so is always treated as plural. We can find data in all situations of the world around us, in all the structured or unstructured, in continuous or discrete conditions, in weather records, stock market logs, in photo albums, music playlists, or in our Twitter account. In fact, data can be seen as the essential raw material to any kind of human activity. According to the Oxford English Dictionary, data are

> *"known facts or things used as basis for inference or reckoning"*.

As it is shown in the following image, we can see data in two distinct ways, **Categorical** and **Numerical**:

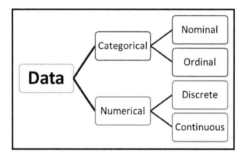

Categorical data are values or observations that can be sorted into groups or categories. There are two types of categorical values, **nominal** and **ordinal**. A nominal variable has no intrinsic ordering to its categories. For example, housing is a categorical variable with two categories (own and rent). An ordinal variable has an established ordering. For example, age as a variable with three orderly categories (young, adult, and elder).

Numerical data are values or observations that can be measured. There are two kinds of numerical values, **discrete** and **continuous**. Discrete data are values or observations can be counted and are distinct and separate, for example, the number of lines in a code. Continuous data are values or observations that may take on any value within a finite or infinite interval, for example, an economic time series like historic gold prices.

The kinds of datasets used in this book are the following:

- E-mails (unstructured, discrete)
- Digital images (unstructured, discrete)
- Stock market logs (structured, continuous)
- Historic gold prices (structured, continuous)
- Credit approval records (structured, discrete)

- Social media friends relationships (unstructured, discrete)
- Tweets and treading topics (unstructured, continuous)
- Sales records (structured, continuous)

For each of the projects in this book we try to use a different kind of data. This book is trying to give the reader the ability to address different kinds of data problems.

The data analysis process

When you have a good understanding of a phenomenon it is possible to make predictions about it. Data analysis helps us to make this possible through exploring the past and creating predictive models.

The data analysis process is composed of following steps:

- The statement of problem
- Collecting your data
- Cleaning the data
- Normalizing the data
- Transforming the data
- Exploratory statistics
- Exploratory visualization
- Predictive modeling
- Validating your model
- Visualizing and interpreting your results
- Deploying your solution

All of these activities can be grouped as is shown in the following image:

The problem

The problem definition starts with high-level business domain questions, such as how to track differences in behavior between groups of customers or knowing what the gold price will be in the next month. Understanding the objectives and requirements from a domain perspective is the key for a successful data analysis project.

Types of data analysis questions include:

- Inferential
- Predictive
- Descriptive
- Exploratory
- Causal
- Correlational

Data preparation

Data preparation is about how to obtain, clean, normalize, and transform the data into an optimal dataset, trying to avoid any possible data quality issues such as invalid, ambiguous, out-of-range, or missing values. This process can take up a lot of time. In Chapter 11, *Working with Twitter Data*, we will go into more detail about working with data, using OpenRefine to address complicated tasks. Analyzing data that has not been carefully prepared can lead you to highly misleading results.

The characteristics of good data are as follows:

- Complete
- Coherent
- Ambiguity elimination
- Countable
- Correct
- Standardized
- Redundancy elimination

Data exploration

Data exploration is essentially looking at the processed data in a graphical or statistical form and trying to find patterns, connections, and relations in the data. Visualization is used to provide overviews in which meaningful patterns may be found. In `Chapter 3`, *Getting to Grips with Visualization,* we will present a JavaScript visualization framework (D3.js) and implement some examples of how to use visualization as a data exploration tool.

Predictive modeling

From the galaxy of information we have to extract usable hidden patterns and trends using relevant algorithms. To extract the future behavior of these hidden patterns, we can use predictive modeling. Predictive modeling is a statistical technique to predict future behavior by analyzing existing information, that is, historical data. We have to use proper statistical models that best forecast the hidden patterns of the data or information.

Predictive modeling is a process used in data analysis to create or choose a statistical model to try to best predict the probability of an outcome. Using predictive modeling, we can assess the future behavior of the customer. For this, we require past performance data of that customer. For example, in the retail sector, predictive analysis can play an important role in getting better profitability. Retailers can store galaxies of historical data. After developing different predicting models using this data, we can forecast to improve promotional planning, optimize sales channels, optimize store areas, and enhance demand planning.

Initially, building predictive models requires expertise views. After building relevant predicting models, we can use them automatically for forecasts. Predicting models give better forecasts when we concentrate on a careful combination of predictors. In fact, if the data size increases, we get more precise prediction results.

In this book we will use a variety of those models, and we can group them into three categories based on their outcomes:

Model	Chapter	Algorithm
Categorical outcome (Classification)	4	Naïve Bayes Classifier
	11	Natural Language Toolkit and Naïve Bayes Classifier
Numerical outcome (Regression)	6	Random walk

	8	Support vector machines
	8	Distance-based approach and k-nearest neighbor
	9	Cellular automata
Descriptive modeling (Clustering)	5	**Fast Dynamic Time Warping** (**FDTW**) + distance metrics
	10	Force layout and Fruchterman-Reingold layout

Another important task we need to accomplish in this step is finishing the evaluating model we chose as optimal for the particular problem.

Model assumptions are important for the quality of the predictions model. Better predictions will result from a model that satisfies its underlying assumptions. However, assumptions can never be fully met in empirical data, and evaluation preferably focuses on the validity of the predictions. The strength of the evidence for validity is usually considered to be stronger.

The **no free lunch** theorem proposed by Wolpert in 1996 said:

> *"No Free Lunch theorems have shown that learning algorithms cannot be universally good".*

But extracting valuable information from the data means the predictive model should be accurate. There are many different tests to determine if the predictive models we create are accurate, meaningful representations that will prove valuable information.

The model evaluation helps us to ensure that our analysis is not overoptimistic or over fitted. In this book we are going to present two different ways of validating the model:

- **Cross-validation**: Here, we divide the data into subsets of equal size and test the predictive model in order to estimate how it is going to perform in practice. We will implement cross-validation in order to validate the robustness of our model as well as evaluate multiple models to identify the best model based on their performance.
- **Hold-out**: Here, a large dataset is arbitrarily divided into three subsets: training set, validation set, and test set.

Visualization of results

This is the final step in our analysis process. When we present model output results, visualization tools can play an important role. The visualization results are an important piece of our technological architecture. As the database is the core of our architecture, various technologies and methods for the visualization of data can be employed.

In an explanatory data analysis process, simple visualization techniques are very useful for discovering patterns, since the human eye plays an important role. Sometimes, we have to generate a three-dimensional plot for finding the visual pattern. But, for getting better visual patterns, we can also use a scatter plot matrix, instead of a three-dimensional plot. In practice, the hypothesis of the study, dimensionality of the feature space, and data all play important roles in ensuring a good visualization technique.

In this book, we will focus in the univariate and multivariate graphical models. Using a variety of visualization tools like bar charts, pie charts, scatterplots, line charts, and multiple line charts, all implemented in D3.js; we will also learn how to use standalone plotting in Python with Matplotlib.

Quantitative versus qualitative data analysis

Quantitative data are numerical measurements expressed in terms of numbers.

Qualitative data are categorical measurements expressed in terms of natural language descriptions.

As is shown in the following image, we can observe the differences between quantitative and qualitative analysis:

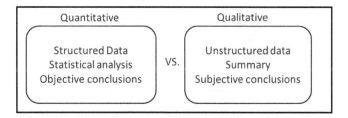

Quantitative analytics involves analysis of numerical data. The type of the analysis will depend on the level of measurement. There are four kinds of measurements:

- Nominal data has no logical order and is used as classification data.
- Ordinal data has a logical order and differences between values are not constant.

- Interval data is continuous and depends on logical order. The data has standardized differences between values, but do not include zero.
- Ratio data is continuous with logical order as well as regular intervals differences between values and may include zero.

Qualitative analysis can explore the complexity and meaning of social phenomena. Data for qualitative study may include written texts (for example, documents or e-mail) and/or audible and visual data (digital images or sounds). In `Chapter 11`, *Working with Twitter Data*, we will present a sentiment analysis from Twitter data as an example of qualitative analysis.

Importance of data visualization

The goal of data visualization is to expose something new about the underlying patterns and relationships contained within the data. The visualization not only needs to be beautiful but also meaningful in order to help organizations make better decisions. Visualization is an easy way to jump into a complex dataset (small or big) to describe and explore the data efficiently. Many kinds of data visualization are available, such as bar charts, histograms, line charts, pie charts, heat maps, frequency Wordles (as is shown in the following image), and so on, for one variable, two variables, many variables in one, and even two or three dimensions:

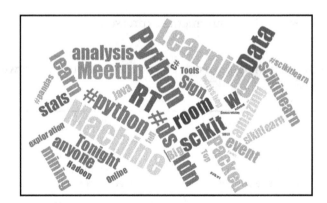

Data visualization is an important part of our data analysis process because it is a fast and easy way to perform exploratory data analysis through summarizing their main characteristics with a visual graph.

The goals of exploratory data analysis are as follows:

- Detection of data errors
- Checking of assumptions
- Finding hidden patters (like tendency)
- Preliminary selection of appropriate models
- Determining relationships between the variables

We will go into more detail about data visualization and exploratory data analysis in `Chapter 3`, *Getting to Grips with Visualization*.

What about big data?

Big data is a term used when the data exceeds the processing capacity of a typical database. The integration of computer technology into science and daily life has enabled the collection of massive volumes of data, such as climate data, website transaction logs, customer data, and credit card records. However, such big datasets cannot be practically managed on a single commodity computer because their sizes are too large to fit in memory, or it takes more time to process the data. To avoid this obstacle, one may have to resort to parallel and distributed architectures, with multicore and cloud computing platforms providing access to hundreds or thousands of processors. For the storing and manipulation of big data, parallel and distributed architectures show new capabilities.

Now, big data is a truth: the variety, volume, and velocity of data coming from the Web, sensors, devices, audio, video, networks, log files, social media, and transactional applications reach exceptional levels. Now big data has also hit the business, government, and science sectors. This phenomenal growth means that not only must we understand big data in order to interpret the information that truly counts, but also the possibilities of big data analytics.

There are three main features of big data:

- **Volume**: Large amounts of data
- **Variety**: Different types of structured, unstructured, and multistructured data
- **Velocity**: Needs to be analyzed quickly

As is shown in the following image, we can see the interaction between these three Vs:

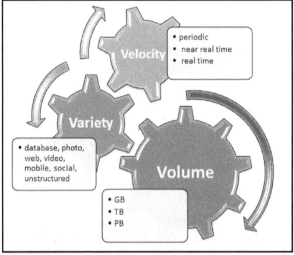

We need big data analytics when data grows fast and needs to uncover hidden patterns, unknown correlations, and other useful information that can be used to make better decisions. With big data analytics, data scientists and others can analyze huge volumes of data that conventional analytics and business intelligence solutions cannot in order to transform business decisions for the future. Big data analytics is a workflow that distils terabytes of low-value data.

Big data is an opportunity for any company to take advantage of data aggregation, data exhaustion, and metadata. This makes big data a useful business analytics tool, but there is a common misunderstanding of what big data actually is.

The most common architecture for big data processing is through Map-Reduce, which is a programming model for processing large datasets in parallel using a distributed cluster.

Apache Hadoop is the most popular implementation of MapReduce, and it is used to solve large-scale distributed data storage, analysis, and retrieval tasks. However, MapReduce is just one of three classes of technologies that store and manage big data. The other two classes are NoSQL and **Massively Parallel Processing** (**MPP**) data stores. In this book we will implement MapReduce functions and NoSQL storage through MongoDB in Chapter 12, *Data Processing and Aggregation with MongoDB*, and Chapter 13, *Working with MapReduce*.

MongoDB provides us with document-oriented storage, high availability, and map/reduce flexible aggregation for data processing.

A paper published by IEEE in 2009 *The Unreasonable Effectiveness of Data* says the following:

> *"But invariably, simple models and a lot of data trump over more elaborate models based on less data."*

This is a fundamental idea in big data (you can find the full paper at `http://static.googl eusercontent.com/media/research.google.com/en//pubs/archive/35179.pdf`). The trouble with real-world data is that the probability of finding false correlations is high and gets higher as the datasets grows. That's why, in this book, we will focus on meaningful data instead of big data.

One of the main challenges for big data is how to store, protect, back up, organize, and catalog the data in a petabyte scale. Another of the main challenges of big data is the concept of data ubiquity. With the proliferation of smart devices with several sensors and cameras, the amount of data available for each person increases every minute. Big data must be able to process all those data in real time:

Quantified self

Quantified self is self-knowledge through self-tracking with technology. In this aspect, one can collect daily activities data on his own in terms of inputs, states, and performance. For example, input means food consumption or quality of surrounding air, states means mood or blood pressure, and performance means mental or physical condition. To collect these data, we can use wearable sensors and life logging. Quantified self-process allows individuals to quantify biometrics that they never knew existed, as well as make data collection cheaper and more convenient. One can track their insulin and cortisol levels and sequence DNA. Using quantified self data, one can be cautious about one's overall health, diet, and level of physical activity.

These days, wearing self-tracking gadgets is rapidly increasing. If we pooled the quantified self-data of a specific group of people, we can apply predictive algorithms on this data to diagnose patients in that location. That means quantified self data is very useful in certain medication contexts.

In the following screenshot, we can see some electronics gadgets that gather quantitative data:

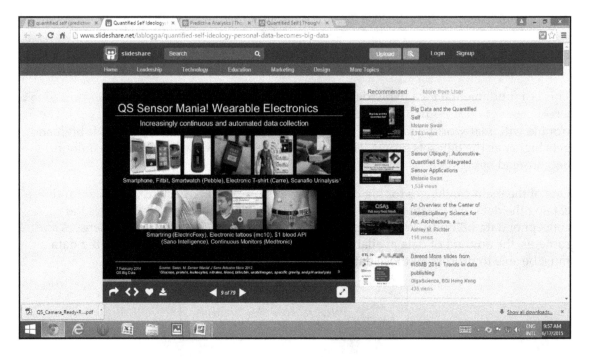

Sensors and cameras

Interaction with the outside world is highly important in data analysis. Using sensors like **Radio-Frequency Identification (RFID)** or a smartphone to scan a **QR** code (**Quick Response)** code are easy ways of interacting directly with the customer, making recommendations, and analyzing consumer trends.

On the other hand, people are using their smartphones all the time, using their cameras as a tool. In Chapter 5, *Similarity-Based Image Retrieval*, we will use these digital images to perform a search by image. This can be used, for example, in face recognition or for finding recommendations of a restaurant just by taking a picture of the front door.

This interaction with the real world can give you a competitive advantage and a real-time data source directly from the customer.

Social network analysis

Nowadays, the Internet brings people together in many ways (that is, using social media); for example, Facebook, Twitter, LinkedIn, and so on. Using these social networks, users are working, playing, socializing online, and demonstrating new forms of collaboration and more. Social networks play a crucial role in reshaping business models and opening up numerous possibilities of studying human interaction and collective behavior.

In fact, if we intended to understand how to identify key individuals in social systems, we can generate models using analytical techniques on social network data and extract the information mentioned previously. This process is called **Social Network Analysis (SNA)**.

Formally, the SNA performs the analysis of social relationships in terms of network theory, with nodes representing individuals and ties representing relationships between the individuals. Social networks create groups of related individuals (friendships) based on different aspects of their interaction. We can find out important information such as hobbies (for product recommendation) or who has the most influential opinion in a group (centrality). We will present in `Chapter 10`, *Working with Social Graphs*, a project, *Who is your closest friend?*, and we will show a solution for Twitter clustering.

Social networks are strongly connected, and these connections are often asymmetric. This makes SNA computationally expensive, and so it needs to be addressed with high-performance solutions that are less statistical and more algorithmic. The visualization of a social network can help us gain a good insight into how people are connected. The exploration of a graph is done through displaying nodes and ties in various colors, sizes, and distributions. D3.js has animation capabilities that enable us to visualize a social graph with interactive animations. These help us to simulate behaviors like information diffusion or the distance between nodes.

Facebook processes more than 500 TB of data daily (images, text, video, likes, and relationships), and this amount of data needs non-conventional treatment like NoSQL databases and MapReduce frameworks. In this book, we will work with MongoDB, a document-based NoSQL database, which also has great functions for aggregations and MapReduce processing.

Tools and toys for this book

The main goal of this book is to provide the reader with self-contained projects ready to deploy, and in order to do this, as you go through the book we will use and implement tools such as Python, D3, and MongoDB. These tools will help you to program and deploy the projects. You also can download all the code from the author's GitHub repository:

```
https://github.com/hmcuesta
```

You can see a detailed installation and setup process of all the tools in `Appendix`, *Setting Up the Infrastructure*.

Why Python?

Python is a "scripting language" – an interpreted language with its own built-in memory management and good facilities for calling and co-operating with other programs. There are two popular versions, 2.7 or 3.x, and in this book we will be focusing on the 3.x version, because this is under active development and has already seen over two years of stable releases.

Python is multi-platform, runs on Windows, Linux/Unix, and Mac OS X, and has been ported to Java and .NET virtual machines. Python has powerful standard libs and a wealth of third-party packages for numerical computation and machine learning, such as NumPy, SciPy, pandas, SciKit, mlpy, and so on.

Python is excellent for beginners, yet great for experts, is highly scalable, and is also suitable for large projects as well as small ones. It is also easily extensible and object-oriented.

Python is widely used by organizations like Google, Yahoo maps, NASA, Red Hat, Raspberry Pi, IBM, and many more.

```
http://wiki.python.org/moin/OrganizationsUsingPython
```

Python has excellent documentation and examples:

```
http://docs.python.org/3/
```

The latest Python software is available for free, even for commercial products, and can be downloaded from here:

```
http://python.org/
```

Why mlpy?

mlpy (**Machine Learning Python**) is a module built on top of NumPy, SciPy, and the GNU scientific libraries. It is open source and supports Python 3.x. mlpy has a large number of machine learning algorithms for supervised and unsupervised problems.

Some of the features of mlpy that will be used in this book are as follows:

- **Regression**: **Support Vector Machines** (**SVM**)
- **Classification**: SVM, **k-nearest-neighbor** (**k-NN**), classification tree
- **Clustering**: k-means, multidimensional scaling
- **Dimensionality Reduction**: **Principal Component Analysis** (**PCA**)
- **Misc**: **Dynamic Time Warping** (**DTW**) distance

We can download the latest version of mlpy from here here:
`http://mlpy.sourceforge.net/`

Reference: D. Albanese, R. Visintainer, S. Merler, S. Riccadonna, G. Jurman, C. Furlanello. *mlpy: Machine Learning Python*, 2012: `http://arxiv.org/abs/122.6548`.

Why D3.js?

D3.js (**data-driven documents**) was developed by Mike Bostock. D3 is a JavaScript library for visualizing data and manipulating the document object model that runs in a browser without a plugin. In D3.js you can manipulate all the elements of the DOM, and it is as flexible as the client-side web technology stack (HTML, CSS, and SVG).

D3.js supports large datasets and includes animation capabilities that make it a really good choice for web visualization.

D3 has excellent documentation, examples and community:

- `https://github.com/mbostock/d3/wiki/Gallery`
- `https://github.com/mbostock/d3/wiki`

We can download the latest version of D3.js from:

`https://d3js.org/`

Why MongoDB?

NoSQL is a term that covers different types of data storage technology that are used when you can't fit your business model into a classical relational data model. NoSQL is mainly used in web 2.0 and in social media applications.

MongoDB is a document-based database. This means that MongoDB stores and organizes the data as a collection of documents. That gives you the possibility to store the view models almost exactly as you model them in the application. You can also perform complex searches for data and elementary data mining with MapReduce.

MongoDB is highly scalable, robust, and works perfectly with JavaScript-based web applications because you can store your data in a JSON document and implement a flexible schema, which makes it perfect for unstructured data.

MongoDB is used by well-known corporations like Foursquare, Craigslist, Firebase, SAP, and Forbes; we can see a detailed list of users at:

```
https://www.mongodb.com/industries
```

MongoDB has a big and active community, as well as well-written documentation:

```
http://docs.mongodb.org/manual/
```

MongoDB is easy to learn and it's free. We can download MongoDB from here:

```
http://www.mongodb.org/downloads
```

Summary

In this chapter, we presented an overview of the data analysis ecosystem and explained the basic concepts of the data analysis process and tools, as well as some insight into the practical applications of data analysis. We have also provided an overview of the different kinds of data, both numerical and categorical. We got into the nature of data: structured (databases, logs, and reports) and unstructured (image collections, social networks, and text mining). Then, we introduced the importance of data visualization and how a fine visualization can help us with exploratory data analysis. Finally, we explored some of the concepts of big data, quantified self-, and social network-analytics.

In the next chapter we will look at the cleaning, processing, and transforming of data using Python and OpenRefine.

2
Preprocessing Data

Building real world data analytic solutions requires accurate data. In this chapter, we discuss how to collect, clean, normalize, and transform raw data into a standard format such as **Comma-Separated Values (CSV)** format or **JavaScript Object Notation (JSON)**, using a tool to process a messy data called **OpenRefine**.

In this chapter, we will cover the following:

- Data sources
- Data scrubbing
- Data reduction methods
- Data formats
- Getting started with OpenRefine

Data sources

Data source is a term for all the technology related to the extraction and storage of data. A data source can be anything from a simple text file to a big database. The raw data can come from observation logs, sensors, transactions, or user behavior.

A **dataset** is a collection of data, usually presented in a tabular form. Each column represents a particular attribute, and each row corresponds to a given member of the data, as is showed in the following screenshot.

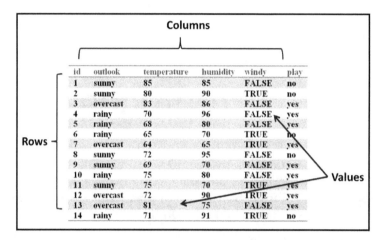

In this section, we will take a look at the most common forms for data sources and datasets.

The data in the preceding screenshot is from the classical *Weather* dataset of the *UC Irvine Machine Learning Repository*: http://archive.ics.uci.edu/ml/

A dataset represents a logical implementation of a data source; the common features of a dataset:

- Dataset characteristics (multivariate and univariate)
- Number of instances
- Area (life, business, and many more)
- Attribute characteristics (real, categorical, and nominal)
- Number of attributes
- Associated tasks (classification or clustering)
- Missing values? (yes or no)

Open data

Open data is data that can be used, reused, and redistributed freely by anyone for any purpose. This is a short list of repositories and databases for open data:

- **Data hub**: http://datahub.io/
- **Book-Crossing Dataset**: http://www.informatik.uni-freiburg.de/~cziegler/BX/
- **World Health Organization**: http://www.who.int/research/en/
- **The World Bank**: http://data.worldbank.org/
- **NASA**: http://data.nasa.gov/
- **United States Government**: http://www.data.gov/
- **Scientific Data from University of Muenster**: http://data.uni-muenster.de/

 Other interesting sources of data come from the data mining and knowledge discovery competitions such as ACM-KDD Cup or Kaggle platform, in most cases, the datasets are still available, even after the competition is closed.

- **ACM-KDD Cup**: http://www.sigkdd.org/kddcup/index.php
- **Kaggle**: http://www.kaggle.com/competitions

Text files

Text files are commonly used for the storage of data because it is easy to transform into different formats and it is often easier to recover and continue processing the remaining contents than with other formats. Big amounts of data come in text format from logs, sensors, e-mails, and transactions. There are several formats for text files such as CSV, **TSV** (tab delimiter), **XML** (**Extensible Markup Language**), and JSON (see section *Data formats*).

Excel files

MS-Excel is probably the most used and also the most underrated data analysis tool. In fact, Excel has some features like filtering, aggregation functions, and using **Visual Basic for Application (VBA)** you can make queries using languages like SQL with the sheets or with an external database:

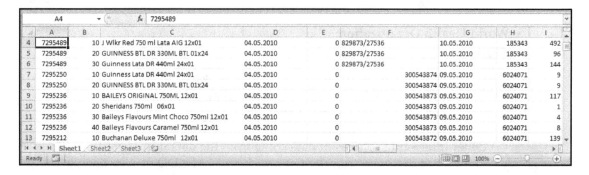

Some Excel disadvantages are that missing values are handled inconsistently and there is no record of how an analysis was accomplished. In the case of the Analysis ToolPak, it can only work with one sheet at a time.

SQL databases

A **database** is an organized collection of data. **Structured Query Language (SQL)** is a database language for managing and manipulating data in **Relational Database Management Systems (RDBMS)**. The **Database Management Systems (DBMS)** is responsible for maintaining the integrity and security of stored data, and for recovering information if the system fails. SQL language is split into two subsets of instructions: the **Data Definition Language (DDL)** and **Data Manipulation Language (DML)**.

The data is organized in **schemas** (database) and divided into **tables** related by logical relationships, where we can retrieve the data by making SQL queries to the main schema, as shown in the following screenshot:

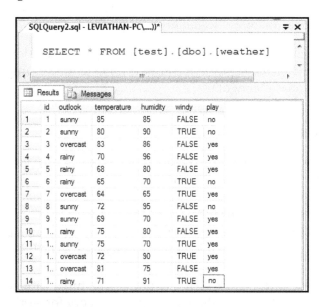

DDL allows us to create, delete, and alter database tables. We can also define keys to specify relationships between tables and implement constraints between database tables:

- CREATE TABLE : This creates a new table
- ALTER TABLE : This alters a table
- DROP TABLE : This one deletes a table

DML is a language which enables users to access and manipulate data:

- SELECT : This command is used to retrieve data from the database
- INSERT INTO : This is used to insert new data into the database
- UPDATE : This is used to modify data in the database
- DELETE : This one is used to delete data in the database

NoSQL databases

Not only SQL (**NoSQL**) is a term used in several technologies where the nature of the data does not require a relational model. NoSQL technologies allow working with a huge quantity of data, higher availability, scalability, and performance.

 See `Chapter 12`, *Data Processing and Aggregation with MongoDB*, and `Chapter 13`, *Working with MapReduce*, for extended examples of document store database MongoDB.

The most common types of NoSQL data stores are:

- **Document store**: Data is stored and organized as a collection of documents. The model schema is flexible and each collection can handle any number of fields. For example, MongoDB uses a document of type BSON (binary format of JSON) and CouchDB uses a JSON document.
- **Key-value store**: Data is stored as key-value pairs without a predefined schema. Values are retrieved by their keys. For example, Apache Cassandra, Dynamo, HBase, and Amazon SimpleDB.
- **Graph-based store**: Data is stored in graph structures with nodes, edges, and properties using the computer science graph theory for storing and retrieving data. This kind of database is excellent to represent social network relationships, for example, Neo4js, InfoGrid, and Horton.

For more information about NoSQL see also:

```
http://nosql-database.org/
```

Multimedia

The increasing number of mobile devices makes a priority of data analysis acquire the ability to extract semantic information from multimedia data sources. Data sources include directly perceivable media such as audio, image, and video. Some of the applications for these kinds of data source are:

- Content-based image
- Content-based video retrieval
- Movie and video classification
- Face recognition
- Speech recognition

- Audio and music classification

In `Chapter 5`, *Similarity-Based Image Retrieval*, we present a similarity-based image search engine using Caltech256, that is an image dataset with over 30,600 images.

Web scraping

When we want to obtain data, a good place to start is on the Web. **Web scraping** refers to an application that processes the HTML of a webpage to extract data for manipulation. Web scraping applications will simulate a person viewing a website with a browser. In the following example, we want to get the current gold price from the website `www.gold.org` as is shown in the following screenshot:

Then, we need to inspect the **Gold Spot Price** element on the website, where we will find the HTML tag:

```
<td class="value" id="spotpriceCellAsk">1,573.85</td>
```

We can observe an ID `spotpriceCellAsk` in the `td` tag; this is the element we will get with the next Python code.

> For this example, we will use the `BeautifulSoup` library version 4. In Linux, we need to open a Terminal and execute the next command to install it from the system package manager:
> `$ apt-get install python-bs4`
> On windows, we need to download the library from:
> `http://crummy.com/software/BeautifulSoup/bs4/download/`
> To install it, just execute in the command line:
> `$ python setup.py install`

First, we need to import the `BeautifulSoup` and `urllib.request` libraries:

```
from bs4 import BeautifulSoup
import urllib.request
from time import sleep
from datetime import datetime
```

Then, we use the function `getGoldPrice` to retrieve the current price from the website; in order to do this, we need to provide the `url` parameter to make the request and read the entire page:

```
req = urllib.request.urlopen(url)
page = req.read()
```

Finally, we use `BeautifulSoup` to parse the page (creating a list of all the elements of the page) and ask for the element `td` with the ID `spotpriceCellAsk`:

```
scraping = BeautifulSoup(page)
price=  scraping.findAll("td",attrs={"id":"spotpriceCellAsk"})[0].text
```

Now, we return the variable `price` with the current gold price, this value changes every minute in the website. In this case, we want all the values in an hour, so we call the function `getGoldPrice` in a `for` loop 60 times, making the script wait 59 seconds for each call:

```
for x in range(0,60):

    sleep(59)
```

Finally, we save the result in a file `goldPrice.out` and include the current date/time in the format `HH:MM:SS[AM or PM]` (11:35:42 PM for example) separated by a comma:

```
with open("goldPrice.out","w") as f:
    ...
        sNow = datetime.now().strftime("%I:%M:%S%p")
        f.write("{0}, {1} \n ".format(sNow, getGoldPrice()))
```

The function `datetime.now().strftime` creates a string representing the time under the control of an explicit format string `"%I:%M:%S%p"` where `%I` represents hour as a decimal number from 0 to 12, `%M` represents minutes as a decimal number from 00 to 59, `%S` represents seconds as a decimal number from 00 to 61, and `%p` represents either AM or PM.

A list of complete format directives can be found at the following link:

```
http://docs.python.org/3.4/library/datetime.html
```

Here is the full script:

```
from bs4 import BeautifulSoup
import urllib.request
from time import sleep
from datetime import datetime
def getGoldPrice():
    url = "http://gold.org"
    req = urllib.request.urlopen(url)
    page = req.read()
    scraping = BeautifulSoup(page)
    price= scraping.findAll
("td",attrs={"id":"spotpriceCellAsk"})[0]
    .text
    return price

with open("goldPrice.out","w") as f:
    for x in range(0,60):
        sNow = datetime.now().strftime("%I:%M:%S%p")
        f.write("{0}, {1} \n ".format(sNow, getGoldPrice()))
        sleep(59)
```

You can download the full script (`WebScraping.py`) from the author's GitHub repository:
https://github.com/hmcuesta/PDA_Book/tree/master/Chapter2

The output file `goldPrice.out` will look like this:

```
11:35:02AM, 1481.25
11:36:03AM, 1481.26
11:37:02AM, 1481.28
11:38:04AM, 1481.25
11:39:03AM, 1481.22
...
```

Data scrubbing

Scrubbing data, also called **data cleansing**, is the process of correcting or removing data in a dataset that is incorrect, inaccurate, incomplete, improperly formatted, or duplicated.

The result of the data analysis process not only depends on the algorithms, it depends on the quality of the data. That's why the next step after obtaining the data, is data scrubbing. In order to avoid dirty data, our dataset should possess the following characteristics:

- Correct
- Completeness
- Accuracy
- Consistency
- Uniformity

Dirty data can be detected by applying some simple statistical data validation and also by parsing the texts or deleting duplicate values. Missing or sparse data can lead you to highly misleading results.

Statistical methods

In this method, we need some context about the problem (knowledge domain) to find values that are unexpected and thus erroneous, even if the data type matches but the values are out of the range. This can be resolved by setting the values to an average or mean value. Statistical validations can be used to handle missing values, which can be replaced by one or more probable values using interpolation or by reducing the dataset using decimation:

- **Mean**: This is the value calculated by summing up all values and then dividing by the number of values.
- **Median**: The median is defined as the value where 50% of values in a range will be below 50% of values above the value.
- **Range constraints**: These are numbers or dates which should fall within a certain range. That is, they have minimum and/or maximum possible values.
- **Clustering**: Usually, when we obtain data directly from the user some values include ambiguity or refer to the same value with a typo. For example, `"Buchanan Deluxe 750ml 12x01"` and `"Buchanan Deluxe 750ml 12x01."` which are different only by a ".". or in the case of `Microsoft` or `MS` instead of `Microsoft Corporation` which refers to the same company and all values are valid. In those cases, grouping can help us to get accurate data and eliminate duplicated data, enabling a faster identification of unique values.

Text parsing

We perform parsing to help us to validate whether a string of data is well formatted and avoid syntax errors.

Regular expression patterns, usually text fields, would have to be validated this way. For example, dates, e-mail, phone numbers, and IP address. **Regex** is a common abbreviation for **regular expression**.

In Python, we will use the re module to implement regular expressions. We can perform text search and pattern validations.

First, we need to import the re module:

```
import re
```

In the following examples, we will implement three of the most common validations (e-mail, IP address, and date format).

- **E-mail validation**:

```
myString = 'From: readers@packt.com (readers email)'
result = re.search('([\w.-]+)@([\w.-]+)', myString)
if result:
    print (result.group(0))
    print (result.group(1))
    print (result.group(2))
```

- Output:

```
>>> readers@packt.com
>>> readers
>>> packt.com
```

- The search() function scans through a string, searching for any location where the Regex matches. The group() function helps us to return the string matched by the Regex. The \w pattern matches any alphanumeric character and is equivalent to the [a-zA-Z0-9_] class.

- **IP address validation**:

```
isIP = re.compile('\d{1,3}\.\d{1,3}\.\d{1,3}\.\d{1,3}')
myString = " Your IP is:  192.168.1.254  "
result = re.findall(isIP,myString)
print(result)
```

- Output:

```
>>> 192.168.1.254
```

- The `findall()` function finds all the substrings where the Regex matches and returns them as a list. The pattern \d matches any decimal digit, and is equivalent to the class [0-9].

- **Date format:**

```
myString = "01/04/2001"
isDate = re.match('[0-1][0-9]\/[0-3][0-9]\/[1-2][0-9]{3}', myString)
if isDate:
    print("valid")
else:
    print("invalid")
```

- Output:

```
>>> 'valid'
```

- The `match()` function finds whether the Regex matches with the string. The pattern implements the class [0-9] in order to parse the date format.

For more information about regular expressions:
http://docs.python.org/3.4/howto/regex.html#regex-howto

Data transformation

Data transformation is usually related to databases and data warehouses where values from a source format are extract, transform, and load in a destination format.

Extract, Transform, and Load (ETL) obtains data from various data sources, performs some transformation functions depending on our data model, and loads the resulting data into the destination.

- Data extraction allows us to obtain data from multiple data sources, such as relational databases, data streaming, text files (JSON, CSV, and XML), and NoSQL databases.
- Data transformation allows us to cleanse, convert, aggregate, merge, replace, validate, format, and split data.

- Data loading allows us to load data into a destination format, such as relational databases, text files (JSON, CSV, XML), and NoSQL databases.

In statistics data, transformation refers to the application of a mathematical function to the dataset or time series points.

Data formats

When we are working with data for human consumption, the easiest way to store it is in text files. In this section, we will present parsing examples of the most common formats such as CSV, JSON, and XML. These examples will be very helpful in the following chapters.

The dataset used for these examples is a list of Pokemon by National Pokedex number, obtained from:
`http://bulbapedia.bulbagarden.net/`
All the scripts and dataset files can be found in the author's GitHub repository:
`https://github.com/hmcuesta/PDA_Book/tree/master/Chapter3`

CSV is a very simple and common open format for table-like data, which can be exported and imported by most of the data analysis tools. CSV is a plain text format; this means that the file is a sequence of characters, with no data that has to be interpreted instead, such as binary numbers.

There are many ways to parse a CSV file from Python, and here we will discuss two:

The first eight records of the CSV file (`pokemon.csv`) look like this:

```
id, typeTwo, name, type
001, Poison, Bulbasaur, Grass
002, Poison, Ivysaur, Grass
003, Poison, Venusaur, Grass
006, Flying, Charizard, Fire
012, Flying, Butterfree, Bug
013, Poison, Weedle, Bug
014, Poison, Kakuna, Bug
015, Poison, Beedrill, Bug
. . .
```

Parsing a CSV file with the CSV module

First, we need to import the `csv` module:

```
import csv
```

Then, we open the file `.csv` and with the `csv.reader(f)` function we parse the file:

```
with open("pokemon.csv") as f:
    data = csv.reader(f)
    #Now we just iterate over the reader

    for line in data:
        print(" id: {0} , typeTwo: {1}, name:  {2}, type: {3}"
              .format(line[0],line[1],line[2],line[3]))
```

Output:

```
[(1, b' Poison', b' Bulbasaur', b' Grass')
 (2, b' Poison', b' Ivysaur', b' Grass')
 (3, b' Poison', b' Venusaur', b' Grass')
 (6, b' Flying', b' Charizard', b' Fire')
    (12, b' Flying', b' Butterfree', b' Bug')
    . . .]
```

Parsing CSV file using NumPy

First, we need to import the NumPy library:

```
import numpy as np
```

NumPy provides us with the `genfromtxt` function, which receives four parameters. First, we need to provide the name of the file `pokemon.csv`. Then, we skip the first line as a header (`skip_header`). Next, we need to specify the data type (`dtype`). Finally, we will define the comma as the delimiter:

```
data = np.genfromtxt("pokemon.csv"
                     ,skip_header=1
                     ,dtype=None
                     ,delimiter=',')
```

Then, just print the result.

```
print(data)
```

Output:

```
id: id , typeTwo: typeTwo, name:   name, type: type
id:  001 , typeTwo:  Poison, name:   Bulbasaur, type:  Grass
id:  002 , typeTwo:  Poison, name:   Ivysaur, type:  Grass
id:  003 , typeTwo:  Poison, name:   Venusaur, type:  Grass
id:  006 , typeTwo:  Flying, name:   Charizard, type:  Fire
 . . .
```

JSON

JSON is a common format to exchange data. Although it is derived from JavaScript, Python provides us with a library to parse JSON.

Parsing JSON file using the JSON module

The first three records of the JSON file (`pokemon.json`) look like this:

```
[
    {
        "id": " 001",
        "typeTwo": " Poison",
        "name": " Bulbasaur",
        "type": " Grass"
    },
    {
        "id": " 002",
        "typeTwo": " Poison",
        "name": " Ivysaur",
        "type": " Grass"
    },
```

```
{
    "id": " 003",
    "typeTwo": " Poison",
    "name": " Venusaur",
    "type": " Grass"
},
. . .]
```

First, we need to import the `json` module and the `pprint` ("pretty-print") module.

```
import json
from pprint import pprint
```

Then, we open the `pokemon.json` file and with the `json.loads` function we parse the file:

```
with open("pokemon.json") as f:
    data = json.loads(f.read())
```

Finally, just print the result with the `pprint` function:

```
pprint(data)
```

Output:

```
    [{'id': ' 001', 'name': ' Bulbasaur', 'type': ' Grass', 'typeTwo': '
Poison'},
    {'id': ' 002', 'name': ' Ivysaur', 'type': ' Grass', 'typeTwo': '
Poison'},
    {'id': ' 003', 'name': ' Venusaur', 'type': ' Grass', 'typeTwo': '
Poison'},
    {'id': ' 006', 'name': ' Charizard', 'type': ' Fire', 'typeTwo': '
Flying'},
    {'id': ' 012', 'name': ' Butterfree', 'type': ' Bug', 'typeTwo': '
Flying'}, . . . ]
```

XML

According to **World Wide Web Consortium (W3C)**:

> *"XML (Extensible Markup Language) is a simple, very flexible text format derived from SGML (ISO 8879). Originally designed to meet the challenges of large-scale electronic publishing, XML is also playing an increasingly important role in the exchange of a wide variety of data on the Web and elsewhere."*

The first three records of the XML file (`pokemon.xml`) look like this:

```
<?xml version="1.0" encoding="UTF-8" ?>
<pokemon>
    <row>
        <id> 001</id>
        <typeTwo> Poison</typeTwo>
        <name> Bulbasaur</name>
        <type> Grass</type>
    </row>
    <row>
        <id> 002</id>
        <typeTwo> Poison</typeTwo>
        <name> Ivysaur</name>
        <type> Grass</type>
    </row>
    <row>
        <id> 003</id>
        <typeTwo> Poison</typeTwo>
        <name> Venusaur</name>
        <type> Grass</type>
    </row>
. . .
</pokemon>
```

Parsing XML in Python using the XML module

First, we need to import the `ElementTree` object from the `xml` module:

```
from xml.etree import ElementTree
```

Then, we open the `pokemon.xml` file and with the `ElementTree.parse` function we parse the file:

```
with open("pokemon.xml") as f:
    doc = ElementTree.parse(f)
```

Finally, just print each `row` element with the `findall` function:

```
for node in doc.findall('row'):
    print("")
    print("id: {0}".format(node.find('id').text))
    print("typeTwo: {0}".format(node.find('typeTwo').text))
    print("name: {0}".format(node.find('name').text))
    print("type: {0}".format(node.find('type').text))
```

Output:

```
id:   001
typeTwo:   Poison
name:   Bulbasaur
type:   Grass
id:   002
typeTwo:   Poison
name:   Ivysaur
type:   Grass
id:   003
typeTwo:   Poison
name:   Venusaur
type:   Grass
.  .  .
```

YAML

YAML (YAML Ain't Markup Language) is a human-friendly data serialization. It's not as popular as JSON or XML, but it was designed to be easily mapped to data types common to most high-level languages. A Python parser implementation called PyYAML is available in the PyPI repository, and its implementation is very similar to the JSON module.

The first three records of the YAML file (`pokemon.yaml`) look like this:

```
Pokemon:
  -id       :  001
typeTwo :  Poison
name      :  Bulbasaur
type      :  Grass
  -id       :  002
typeTwo :  Poison
name      :  Ivysaur
type      :  Grass
  -id       :  003
typeTwo :  Poison
name      :  Venusaur
type      :  Grass
.  .  .
```

Data reduction methods

Many data scientists use large data size in volume for analysis, which takes a long time, though it is very difficult to analyze the data sometimes. In data analytics applications, if you use a large amount of data, it may produce redundant results. In order to overcome such difficulties, we can use data reduction methods.

Data reduction is the transformation of numerical or alphabetical digital information derived empirically or experimentally into a corrected, ordered, and simplified form. Reduced data size is very small in volume and comparatively original, hence, the storage efficiency will increase and at the same time we can minimize the data handling costs and will minimize the analysis time also.

We can use several types of data reduction methods, which are listed as follows:

- Filtering and sampling
- Binned algorithm
- Dimensionality reduction

Filtering and sampling

In data reduction methods, filtering plays an important role. Filtering explains the process of detecting and correcting errors from raw data. After getting the filtered data, we can use them as input data for succeeding analysis. The filters look like mathematical formulas. There are many filtering methods available for extracting errors and noise-free data from raw data. Some of the filter methods are moving average filtering, Savitzky-Golay filtering, high correlation filtering, Bayesian filtering, and many more. These filters should be adopted in a proper manner based on raw data and the context of the study.

Most of the filters are applied to a sample of raw data. For example, in Bayesian filtering methods, we can use a sample of data, that comes based on the Monte Carlo sequential sampling method.

For data reduction using filtering, the sampling techniques play an important role. The importance of sampling is to extract statistical inferences about the population from sample data. Large data stored in the databases is normally called "population" data. In the data reduction process, we extract a subset of data that best represents the population.

Binned algorithm

Binning is a classification process for extracting a small set of groups or bins from a continuous variable. Binning is widely used in many fields like genomics and credit scoring. More frequently, binning is used at an early stage to select variables from the specified fields. To enhance the predictive power, similar attributes of an independent variable are grouped into the same bin.

Commonly used binning algorithms are:

- **Equal-width binning**: Values are divided into a predefined number of bins of equal width intervals.
- **Equal-size binning**: Attributes are sorted first and then divided into a pre-defined number of equal-size bins.
- **Optima binning**: Data divided into a large number of initial equal-width bins, say 20. These bins are then treated as categories of a nominal variable and grouped to the required number of segments in a tree structure.
- **Multi-interval discretization binning**: This binning process is the minimization of entropy for binary discretizing the range of a continuous variable into multiple intervals and recursively defining the best bins.

For selecting proper binning algorithm, we should consider the following strategies:

- Missing values are binned separately
- Each bin should contain at least 5% of observations
- No bins have 0 accounts for good or bad
- **Weight of Evidence (WOE)** is a quantitative method for combining evidence in support of a statistical hypothesis
- Binning algorithms are available in Python 3.4 software also, for example, `import binningx0dt` command

Dimensionality reduction

Dimensionality reduction methods should be applied for taking a fewer sequence of orthogonal latent variables, instead of observed explanatory variables as dependent variables in the predicting process. This means dimensionality reduction is converting very high-dimensionality data into much lower dimensionality data, such that each of the lower dimension data conveys more information.

Dimensionality reduction process is a statistical or mathematical technique, in which we can describe most, but not all, of the variance within our data, but retain the relevant information. In statistics, the process of dimension reduction can reduce the number of random variables and can be divided into feature selection and feature extraction. The following diagram represents the process of data reduction:

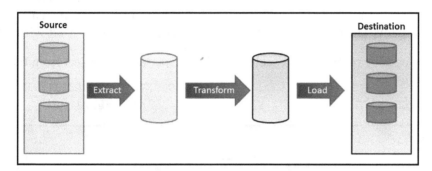

There are many techniques available to tackle dimensionality reduction. **Principle Component Analysis (PCA)** and **Linear Discriminant Analysis (LDA)** are the most widely used techniques. Comparatively, LDA will give better results than PCA for big datasets.

PCA is a multivariate data analysis technique. Using this technique, we can explain the underlying variance-covariance structure of a large set of variables through a few linear combinations of these variables.

The objective of LDA is to perform dimensionality reduction while preserving as much of the class discriminatory information as possible. LDA finds most discriminant projection by maximizing between-class distance and minimizing within-class distance.

Getting started with OpenRefine

OpenRefine (former Google Refine) is a formatting tool very useful in data cleansing, data exploration, and data transformation. It is an open source web application which runs directly on your computer skipping the problem of uploading delicate information to an external server.

To start working with OpenRefine just run the application and open a browser typing the URL:

```
http://127.0.0.1:3333/
```

See the section: Installation of OpenRefine in `Appendix`, *Setting Up the Infrastructure*.

First, we need to upload our data and click on **Create Project**. In the following screenshot, we can observe our dataset; in this case, we will use the monthly sales of an alcoholic beverages company. The dataset format is an MS Excel (`.xlsx`) worksheet with 160 rows.

We can download the original MS Excel file and the OpenRefine project from the author's GitHub repository:

`https://github.com/hmcuesta/PDA_Book/tree/master/Chapter2/`

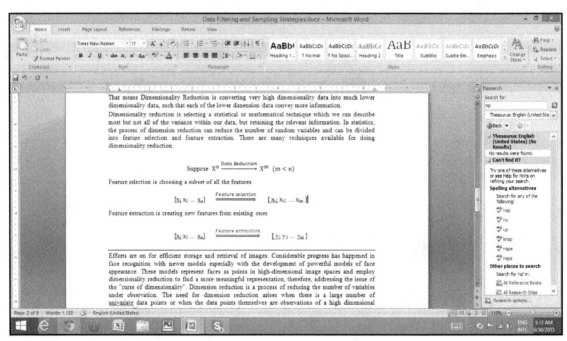

Text facet

Text Facet is a very useful tool, similar to the filter in a spreadsheet. Text facet groups unique text values into groups. This can help us to merge information, and we can see values which could be spelled in a lot of different ways.

Now, we will create a text facet on the **name** column by clicking on that column's drop-down menu and selecting **Facet | Text Facet**. In the following screenshot, we can see the column **name** is grouped by its content. This is helpful to see the distribution of elements in the dataset. We will observe the number of choices (43 in this example), and we can sort the information by name or by count:

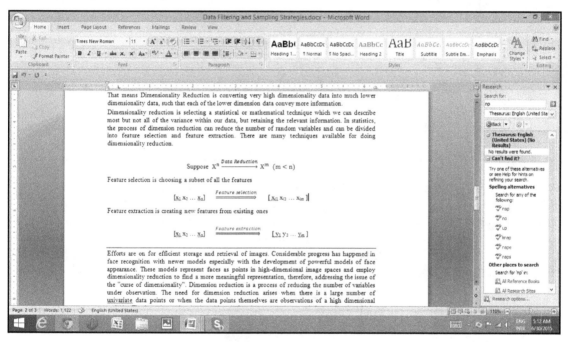

Clustering

Clicking in our text facet (see the following screenshot), we can cluster all the similar values; in this case, we find **Guinness Lata DR 440ml 24×01** and **Guinness Lata DR 440ml 24×01**. Obviously, the " . " in the second value is a typo. The option **Cluster** allows finding this kind of dirty data easy. Now, we just select the option **Merge?** and define the new cell value; we click on **Merge Selected & Close**. Refer to the following screenshot:

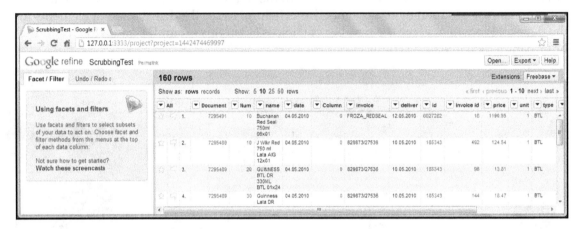

We can play with the parameters of the **Cluster** option like changing the method from **Key collision** to **nearest neighbors**, selecting the rows in a cluster or the length variance of choices. Playing with the parameters, we can find duplicate items in a data column and more complex misspells such as in the following screenshot, where the values **JW Black Label 750ml 12×01** and **JW Bck Label 750ml 12×01** refer to the same product with a typo in the color:

Text filters

We may filter a column using a specific text string or using a regular expression (Java's regular expressions). We will click on the option **Find** of the column we want to filter and then type our search string in the textbox to the left. For more information about Java's regular expressions see:

```
http://docs.oracle.com/javase/tutorial/essential/regex/
```

Numeric facets

Numeric facet groups number into numeric range bins. You can customize numeric facets much like you can customize text facets. For example, if the numeric values in a column are drawn from a power law distribution (see **1** in the following screenshot), then it's better to group them by their logs (see **2** in the same screenshot) using the `value.log()` expression:

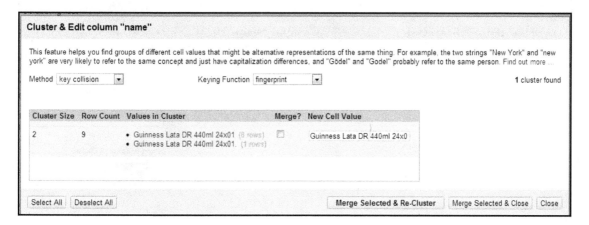

If our values are periodic, we could take the modules by the period to find a pattern, using this expression:

```
mod(value, 6)
```

We can create a numeric facet from a text by taking the length of the string, using this expression:

```
value.length()
```

Transforming data

In our example, the column **date** uses a special date format `01.04.2013`, and we want to replace the "`.`" with "`/`". Fixing this is pretty easy using **Transform**. We need to select the column **date** | **Edit Cells** | **Transform**.

We are going to write a **replace()** expression like this:

```
replace(value,".","/")
```

Now, just click on the **OK** button to apply the transformation:

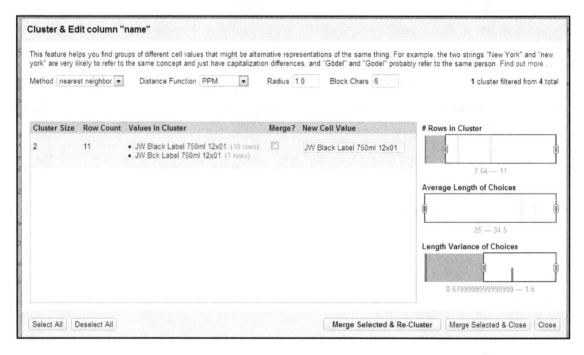

Google Refine Expression Language (GREL) allows us to create complex validations. For example, for a simple business logic when the column value reaches 10 units we make a discount of 5%, we do this with an `if()` statement and some algebra:

```
if(value>10,value*.95,value)
```

 See the following link for a complete list of functions supported by GREL:
`https://code.google.com/p/google-refine/wiki/GRELFunctions`

Exporting data

We can export data from an existing OpenRefine project in several formats like:

- **Tab Separated Values (TSV)**
- CSV
- Excel
- HTML table

To export as a JSON, we need to select the option **Export** and **Templating Export**, where we can specify a JSON template as is shown in the following screenshot:

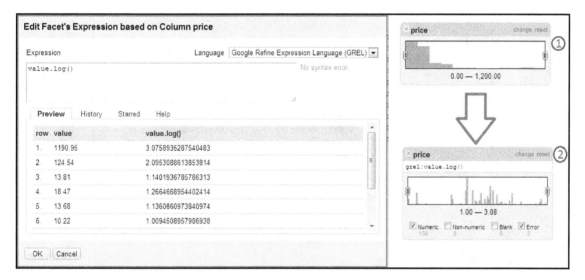

Operation history

We can save all the transformations applied to our dataset just by clicking on the tab **Undo/Redo** and then selecting **Extract**. This will show all the transformations applied to the current dataset (see the following screenshot). Finally, we will copy the generated JSON and it in a text file.

To apply the transformations to another dataset, we just need to open the dataset in OpenRefine and then go to the **Undo/Redo** tab. Click on the **Apply** button and copy the JSON from the first project:

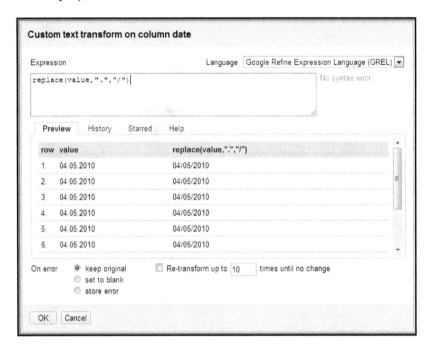

Summary

In this chapter, we explored the common data sources and implemented a web scraping example. Next, we introduced the basic concepts of data scrubbing such as statistical methods and text parsing. Then, we learned about how to parse the most used text formats with Python. We discussed different data reduction methods. Finally, we presented an introduction to OpenRefine, which is an excellent tool for data cleansing and data formatting.

Working with data is not only code or clicks, we need to understand the business domain and play with the data and follow our intuition to get our data in great shape. We need to get involved in the knowledge domain of our data to find inconsistencies. The overall idea of data helps us to discover what we need to know about our data.

In the next chapter, we will explore our data through some visualization techniques and we will present a fast introduction to D3.js.

3

Getting to Grips with Visualization

Sometimes, we don't know how valuable data is until we look at it. In this chapter, we will look into a JavaScript-based web visualization framework called **D3 (Data-Driven Documents)** to create visualizations that make complex information easier to understand. We will cover the following topics:

- What is visualization?
- The visualization lifecycle
- Visualizing different types of data
- Data from social networks
- An overview of visualization analytics

Exploratory Data Analysis (EDA), as mentioned in Chapter 2, *Preprocessing Data*, is a critical part of the data analysis process because it helps us to detect mistakes, determinate relationships, and tendencies, identify outliers, trends, and patterns, or check assumptions. In this chapter, we will present some examples of visualization methods for EDA with discrete and continuous data.

The four types of EDA are **univariate nongraphical**, **multivariate nongraphical**, **univariate graphical**, and **multivariate graphical**. The nongraphical methods refer to the calculation of summary statistics or the outlier detection. In this book, we will focus on the univariate and multivariate graphical models, using a variety of visualization tools such as bar charts, pie charts, scatter plots, line charts, and multiple line charts, all implemented in D3.js.

In this chapter, we will work with two types of data: discrete data with a list of summarized Pokemon types (see Chapter 2, *Preprocessing Data*), and, on the other hand, continuous data using historical exchange rates from March 2008 to March 2013. We will also explore the creation of a random dataset.

What is visualization?

Visualization is a perfect technique for creating diagrams, images, or animations to understand and communicate a message for specific purposes. Literally, visualization is a process of mapping information to visuals. We can communicate both abstract and concrete ideas using visualization through effective visual imagery.

As the volume and complexity of data produced in the arts, engineering, science, education, medicine, and interactive multimedia are growing, so too has the application of visualization in order to understand the content of the data. Presently, the field of computer graphics is found to be important in visualization application. Just as the invention of computer graphics was an important development in visualization, similarly, the development of animation has also helped advance visualization in a greater manner.

Good visualizations present a visual interpretation of the data and also improve decision making. For exploratory analysis, analytics, and for presentation of results, we can apply data visualization techniques extensively. Thus, data visualization is exploding in popularity in various fields. We will discuss web-based visualization with examples in this chapter in a comprehensive manner.

Working with web-based visualization

A data journalist and information designer David McCandless stated in his TED talk:

> *"By visualizing information, we turn it into a landscape that you can explore with your eyes, a sort of information map. And when you're lost in information, an information map is kind of useful."*

The **World Wide Web** (**WWW**) is an information space, which publishes and displays information via the Internet. The accessibility of information through online information services and the WWW is increasing day by day. A vast amount of information is available on websites, which can be accessed through the Internet. But the user may find the information difficult to read, and hence spends a lot of time on the web. Therefore, on websites, the information should be user friendly and reading through it should not take too much time.

Most people prefer pictures over text, so the use of graphical representation of information is increasing daily, leading us to an understanding of how information is structured. We can create and post graphs for the visualization of information in a highly interactive manner all over the web.

To understand the concept well, let's look at the following screenshot that was created and posted interactively on a website:

In this chapter, we will discuss and use D3, the web-based visualization tool.

Exploring scientific visualization

A picture is worth a thousand words enunciates the idea that complex information can be described with just a single image. In the context of scientific visualization, the idea can be applicable, too. The visual representations of complex data sets from simulations, scientific experiments, medical scanners, etc., are called scientific visualization, which is a well-known technique from the areas of interactive computer graphics signal processing, image processing, and system design.

In understanding and solving scientific problems, scientific visualization unites mathematical models and computer graphics from the physical world and creates a visual framework. The algorithmic and mathematical foundations of scientific visualization can be explained by the real-world data from scientific and biomedical domains. In most cases, the application of scientific visualization techniques has been found within the domains of earth science, medical science, physical science, chemical science, applied mathematics, and computer science.

Visualization in art

Visual artists have been using computer-based tools in visual arts rather than traditional art media more often. Artists often use computers for capturing or creating images and forms, editing those images and forms, and printing the images. In turn, they integrate digital technologies and algorithmic art with many traditional arts, and as a result, new media visuals are created, but they are a common theme between the one who prepares the scientific visualization and the artist.

On the other hand, computer usage has blurred the differences between photographers, illustrators, photo editors, handicraft artists, and 3-D modelers. Sophisticated visualization tools are required to reduce the differences between them. This leads handicrafts to become computer-aided imagery, photographers to become digital artists, and illustrators to become animators. While solving visual artists' professional tasks, the visualization processes and experiences of visual artists are unique from those of humanities professionals and scientists at all stages of imagery processing.

The visualization life cycle

The importance and existence of data visualization should be discussed in a visualization lifecycle. There are different steps involved in a visualization lifecycle, these are source of the data, database, data filtering, and generating visuals:

- **Source of the data**: A huge volume of data can be generated from different sources, such as retail stores, social media, financial sectors, and so on.
- **Database**: We can dump the data in databases. There are different database types available, such as XML, JSON, CSV, MY SQL, and so on.

- **Data filtering**: First, we have to check the data type, the structure of the data, and how many columns the data stores. Then, we can go for data filtering using parsing, filtering, aggregation, and cleansing, and for normalization methods if required, to get structured data. Afterward, we can generate a common output format for visualization tools (e.g. CSV).

After getting the structured data or contextual data, we can apply visualization tools to generate the required visuals.

The following diagram explains the visualization lifecycle:

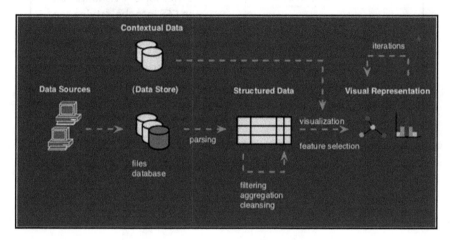

Visualizing different types of data

D3 is a project featured by the Stanford Visualization Group developed by Mike Bostock.

D3 provides us with web-based visualization, which is an excellent way to deploy information and illustrate things like proportions, relationships, correlations, patterns, and to discover things previously unknown. A web browser provides us with a very flexible and interactive interface on practically any device, such as a PC, tablet, or smartphone. D3 is an amazing tool for visualization based on data using **HTML**, **JavaScript**, **SVG**, and **CSS**.

In Chapter 1, *Getting Started* we talked about the importance of data visualization, but in this chapter, we will present examples for you to understand the use of D3.js. In the following screenshot, we can see the basic structure of an HTML document. D3 is going to be included in a basic script tag or into a JavaScript file (.js).

```
1     <!DOCTYPE html>
2     <html>
3       <head>
4         <title> HTML Hello World </title>
5         <style>
6         body {
7           font: 10px arial;
8         }
9         </style>
10      </head>
11      <body>
12        <p> My first paragraph! </p>
13        <script>

15            D3 + JavaScript Code
16            . . .
17
18        </script>
19
20      </body>
21    </html>
22
```

We will define the basic styles for our visualization in **CSS** into the **<style>** tag.

The **D3** code is written in **JavaScript** and will be contained in the **<script>** tag into the body of the **HTML** file

HTML

Hyper Text Markup Language (**HTML**) provides the basic skeleton for our visualization. An HTML document will define the structure of our web page based on a series of tags, which are labels inside angled brackets (<br/>) commonly occurring in pairs (<p>...</p>). D3 will take advantage of the structure of HTML by creating new elements in the document structure, such as by adding new div tags (which define a section in a document). We can see the basic structure of an HTML document in the previous screenshot.

For a complete reference on HTML, please refer to the following link: http://www.w3schools.com/html/

DOM

A **Document Object Model** helps us to represent and interact with objects in HTML documents. Objects in the DOM tree can be addressed and manipulated by programming languages like Python or JavaScript through the elements (tags) of the web page. D3 will change the structure of the HTML document by accessing the DOM tree either by the element ID or its type.

CSS

Cascading Style Sheets can help us to style the web page. A CSS style is based on rules and selectors. We can apply styles to a specific element (tag) through selectors. An example of CSS follows:

```
<style>
body {
   font: 10px arial;
}
</style>
```

JavaScript

JavaScript is a dynamic scripting programming language typically implemented in the client (web browser). All the code in D3.js is developed with JavaScript. JavaScript will help us to create great visualizations with full interactivity which can be updated in real time. In D3.js, we can link to the library (which is stored in a separate file) directly with the snippet listed here:

```
<script src="http://d3js.org/d3.v3.min.js"></script>
```

SVG

Scalable Vector Graphics is an XML-based vector image format for two-dimensional graphics. SVG can be directly included in your web page. SVG provides basic shape elements like rectangle, line, circle, and text to build complicated lines and shapes inside a canvas. Much of the success of D3 is due to the implementation of a wrapper for SVG. With D3, we will not have to modify the XML directly; instead, D3 provides an API to help us place our elements (rectangle, circle, line, etc.) in the correct location on the canvas.

Getting started with D3.js

First, download the latest version of D3 from the official website at `http://d3js.org/`.

To go directly to the latest release, copy this snippet:

```
<script src="http://d3js.org/d3.v3.min.js"></script>
```

In our basic examples, we can just open our HTML document in a web browser to view it. But when we need to load external data sources, we need to publish the folder on a web server like Apache, nginx, or IIS. Python provides us with an easy way to run a web server with `http.server`, so we just need to open the folder where our D3 files are present and execute the following command in the terminal:

```
$ python3 -m http.server 8000
```

In Windows, you can use the same command by removing the number 3 from Python:

```
> python -m http.server 8000
```

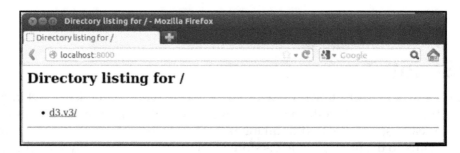

The following examples are based on the Mike Bostock reference gallery, which can be found at `https://github.com/mbostock/d3/wiki/Gallery`.

All the codes and datasets of this chapter may be found in the author's GitHub repository at `https://github.com/hmcuesta/PDA_Book/tree/master/Chapter3`.

Bar chart

The most common visualization tool is probably the bar chart. As we can see in the following diagram, the horizontal axis (**X**) represents the category data and the vertical axis (**Y**) represents a discrete value. We can see the count of Pokemon instances by type with a random sort.

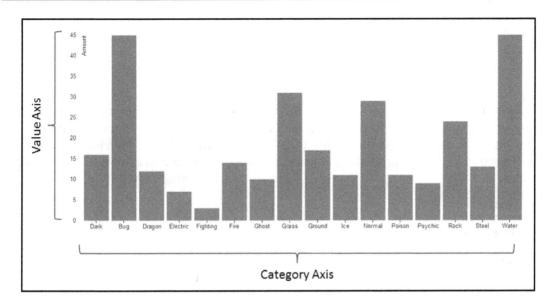

Discrete data are values that can only take certain values, in this case, the number of Pokemon by type.

In the following example, we process the Pokemon list in the JSON format (see Chapter 2, *Preprocessing Data*) and we get the sum of Pokemon by type, sorted by number ascendant, and we then save the result in a CSV format. After data processing, we can visualize the result in a bar-chart.

The first three records of the JSON file (pokemon.json) look like this:

```
[
    {
        "id": " 001",
        "typeTwo": " Poison",
        "name": " Bulbasaur",
        "type": " Grass"
    },
    {
        "id": " 002",
        "typeTwo": " Poison",
        "name": " Ivysaur",
        "type": " Grass"
    },
    {
        "id": " 003",
```

```
            "typeTwo": " Poison",
            "name": " Venusaur",
            "type": " Grass"
        },
    . . . ]
```

In this preprocessing stage, we will use Python to turn the JSON file into a CSV format. We will perform an aggregation to get the number of each category of Pokemon sorted in ascending order. After we get the resulted CSV file, we will start with the visualization in D3.js. The code for the preprocessing is listed as follows. We need to import the necessary modules.

```python
import json
import csv
from pprint import pprint
#Now, we define a dictionary to store the result
typePokemon = {}
#Open and load the JSON file.
with open("pokemon.json") as f:
    data = json.loads(f.read())

#Fill the typePokemon dictionary with sum of Pokemon by type
    for line in data:
        if line["type"] not in typePokemon:
            typePokemon[line["type"]] = 1
        else:
            typePokemon[line["type"]]=typePokemon.get(line["type"])+1

#Open in a write mode the sumPokemon.csv file
with open("sumPokemon.csv", "w") as a:
    w = csv.writer(a)

#Sort the dictionary by number of pokemon
#writes the result (type and amount) into the csv file
    for key, value in sorted(typePokemon.items(),
    key=lambda x: x[1]):
        w.writerow([key,str(value)])
 #finally, we use "pretty print" to print the dictionary
    pprint(typePokemon)
```

The result of the preprocessing can be seen in the following table. Each row has two values: the type and the amount of Pokemon of that type.

Type	Amount
Fighting	3
Electric	7
Psychic	9
Ghost	10
Poison	11
Ice	11
Dragon	12
Steel	13
Fire	14
Dark	16
Ground	17
Rock	24
Normal	29
Grass	31
Water	45
Bug	45

To start working on D3, we need to create a new HTML file with the basic structure (head, style, body). Next, we will include the styles and the script section, as is shown in the following steps.

In the CSS, we specified the style for the axis line, the font family and size for the body, and the bar color.

```
<style>
body {
    font: 14px sans-serif;
}
.axis path,
.axis line {
```

```
    fill: none;
    stroke: #000;
    shape-rendering: crispEdges;
}
.x.axis path {
    display: none;
}
.bar {
    fill: #0489B1;
}
</style>
```

We may define the colors in CSS using the hexadecimal code, like #0489B1, instead of the literal name "blue". The following link is an example of a color selector:
`http://www.w3schools.com/tags/ref_colorpicker.asp`

Inside the body tag we need to refer to the library:

```
<body>
<script src="http://d3js.org/d3.v3.min.js"></script>
```

The first thing we need to do is define a new SVG canvas (`<svg>`) with a `width` and `height` of `1000 x 500` pixels inside the body section of our HTML document.

```
var svg = d3.select("body").append("svg")
    .attr("width", 1000)
    .attr("height", 500)
  .append("g")
    .attr("transform", "translate(50,20)");
```

The `transform` attribute will help us to translate, rotate, and scale a group element (`"g"`). In this case, we want to translate (move) the position of the margins (left and top) on the canvas with translate ("left", "top"). We need to do this because we will need space for the labels in the X and Y axis of our visualization.

Now, we need to open the `sumPokemon.csv` file and read the values from it. Then, we need to create the variable `data` with two attributes, `type`, and `amount`, according to the structure of the CSV file.

The `d3.csv` method will perform an asynchronous request. When the data is available a callback function will be invoked. In this case, we will iterate the list `data` and we will convert the `amount` column to a number as shown in the following snippet:

```
(d.amount = +d.amount).d3.csv("sumPokemon.csv", function(error, data) {
    data.forEach(function(d) {
```

```
    d.amount = +d.amount;
});
```

Now, we will set a labeled **X-Axis** (x.domain) using the map function to get all the type names of the Pokemon. Next, we will use the d3.max function to return the maximum value of each type of Pokemon for the **Y-Axis** as shown in the following snippet:

```
(y.domain). x.domain(data.map(function(d) { return d.type; }));
y.domain([0, d3.max(data, function(d) { return d.amount; })]);
```

Now we will create an SVG group element, which is used to group SVG elements together with the tag <g>. Then, we will use the transform function to define a new coordinate system for a set of SVG elements by applying a transformation to each coordinate specified in this set of SVG elements.

```
svg.append("g")
    .attr("class", "x axis")
    .attr("transform", "translate(0,550)")
    .call(xAxis);

svg.append("g")
    .attr("class", "y axis")
    .call(yAxis)
  .append("text")
    .attr("transform", "rotate(-90)")
    .attr("y", 6)
    .attr("dy", ".71em")
    .style("text-anchor", "end")
    .text("Amount");
```

Finally, we need to generate .bar elements and add them to SVG, then with the data(data) function, for each value in the data we will call the .enter() function and add a "rect" element. D3 allows us to select groups of elements for manipulation through the selectAll function.

> In the following link you can find more information on selections:
> https://github.com/mbostock/d3/wiki/Selections

```
....  svg.selectAll(".bar")
        .data(data)
      .enter().append("rect")
        .attr("class", "bar")
        .attr("x", function(d) { return x(d.type); })
        .attr("width", x.rangeBand())
```

```
      .attr("y", function(d) { return y(d.amount); })
      .attr("height", function(d) { return height - y(d.amount); });

  }); // close the block d3.csv
```

In order to see the result of our visualization, visit
`http://localhost:8000/bar-char.html`; the result is shown in the following diagram:

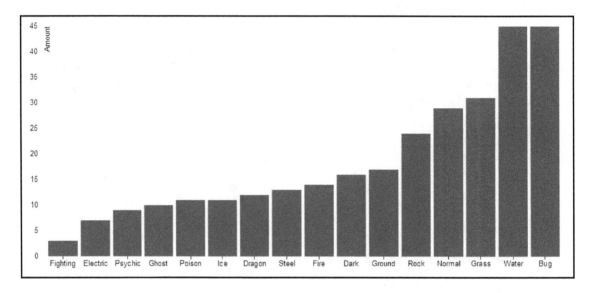

Pie chart

The purpose of the pie chart is to communicate proportions. The sum of all wedges
represents a whole (100 percent). Pie charts help us to understand the distribution of
proportions in an easy way. In this example, we will use the unordered list of Pokemon by
type `sumPokemon.csv`, which can be found at:
`https://github.com/hmcuesta/PDA_Book/tree/master/Chapter3`

We need to define the font family and size for the labels:

```
<style>
body {
  font: 16px arial;
}
</style>
```

Inside the `body` tag we need to refer to the library:

```
<body>
<script src="http://d3js.org/d3.v3.min.js"></script>
```

First, we define the size (width, height, and radius) of the work area:

```
var w = 1160,
    h = 700,
    radius = Math.min(w, h) / 2;
```

Now, we set a range of color that will be used in the chart:

```
var color = d3.scale.ordinal()
    .range(["#04B486", "#F2F2F2", "#F5F6CE", "#00BFFF"]);
```

The function `d3.svg.arc()` creates a circle with an outer radius and an inner radius. See the following pie charts:

```
var arc = d3.svg.arc()
    .outerRadius(radius - 10)
    .innerRadius(0);
```

The function `d3.layout.pie()` specifies how to extract a value from the associated data:

```
var pie = d3.layout.pie()
    .sort(null)
    .value(function(d) { return d.amount; });
```

Now, we select the element body and create a new element, `<svg>`:

```
var svg = d3.select("body").append("svg")
    .attr("width", w)
    .attr("height", h)
  .append("g")
    .attr("transform", "translate(" + w / 2 + "," + h / 2 + ")");
```

Next, we need to open the file "`sumPokemon.csv`", read the values, and create the variable `data` with two attributes: `type` and `amount` as shown in the following snippet:

```
d3.csv("sumPokemon.csv", function(error, data) {
  data.forEach(function(d) {
    d.amount = +d.amount;
  });
```

Finally, we need to generate `.arc` elements and add them to SVG, then with the `data(pie(data))` function, for each value in the data we will call the `.enter()` function and add a "g" element:

```
var g = svg.selectAll(".arc")
    .data(pie(data))
  .enter().append("g")
    .attr("class", "arc");
```

Now we need to apply the style, color, and labels to the group "g" as shown in the following snippet:

```
g.append("path")
    .attr("d", arc)
    .style("fill", function(d) { return color(d.data.type); });
g.append("text")
    .attr("transform", function(d) { return "translate(" +
arc.centroid(d) + ")"; })
    .attr("dy", ".60em")
    .style("text-anchor", "middle")
    .text(function(d) { return d.data.type; });
}); // close the block d3.csv
```

In order to see the result of our visualization, we need to visit the following URL: `http://localhost:8000/pie-char.html`. The result is also shown in the following diagram:

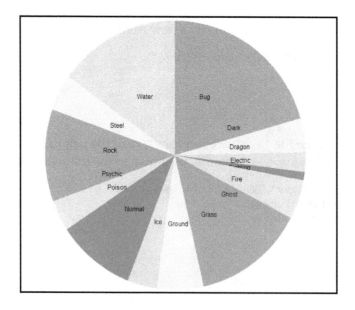

In the following diagram we can see the pie chart with a variation in the attribute's inner radius of 200 pixels in the function `arc`:

```
var arc = d3.svg.arc()
    .outerRadius(radius - 10)
    .innerRadius(200);
```

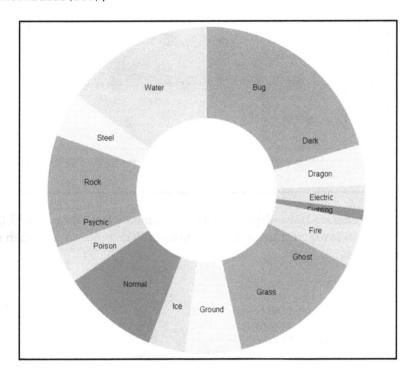

Scatter plots

A **scatter plot** is a visualization tool based in Cartesian space with coordinates of axes **X** and **Y** between two different variables; in this case, it can be value, categorical, or time represented in data points. A scatter plot allows us to see relationships between the two variables.

In the following diagram, we can see a scatter plot, where each has two coordinates: **X** and **Y**. The horizontal axis can take category or time values, and in the vertical axis, we represent a value.

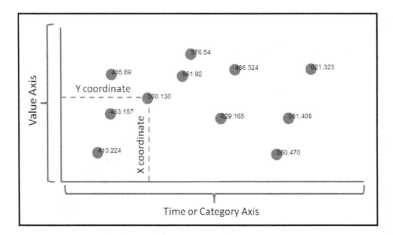

In this example, we generate 20 random points (constrained within a range of 700×500) in a bi-dimensional array in JavaScript using the function `Math.random()` and then store the result in the variable, `data`:

```
var data = [];
for(var i=0; i < 20; i++ ){
    var axisX = Math.round(Math.random() * 700);
    var axisY = Math.round(Math.random() * 500);
    data.push([axisX,axisY]);
}
```

Now we select the element body and create a new element `<svg>` and define its size:

```
var svg = d3.select("body")
        .append("svg")
        .attr("width", 700)
        .attr("height", 500);
```

We use the selector to create a circle for each data point in the variable `data`, defining the coordinate **X** `"cx"` and the coordinate **Y** `"cy"`; define the radius `"r"` to `10` pixels and pick a color, `"fill"`:

```
svg.selectAll("circle")
    .data(data)
    .enter()
    .append("circle")
    .attr("cx", function(d) {      return d[0]; })
```

```
.attr("cy", function(d) {        return d[1]; })
.attr("r", function(d) {        return 10;   })
.attr("fill", "#0489B1");
```

Finally, we create the label of each point including the value of **X**, **Y** coordinates in a text format. We will select a font family, color, and size as shown in the following code snippet:

```
svg.selectAll("text")
    .data(data)
    .enter()
    .append("text")
    .text(function(d) {return d[0] + "," + d[1];  })
    .attr("x", function(d) {return d[0];  })
    .attr("y", function(d) {return d[1];  })
    .attr("font-family", "arial")
    .attr("font-size", "11px")
    .attr("fill", "#000000");
```

In the following screenshot we can see the scatter plot in our web browser:

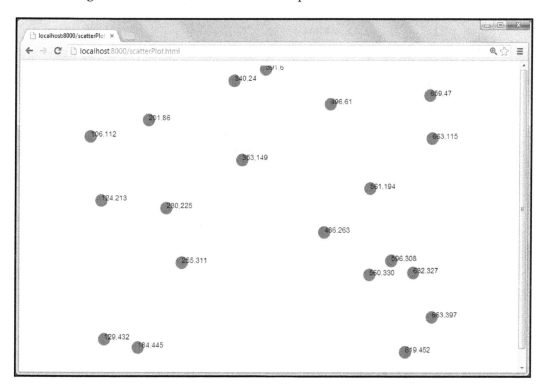

Single line chart

A line chart is a visualization tool that displays continuous data as a series of points connected by a straight line. It is similar to a scatter plot, but in this case, the points have a logical order and the points are connected and are often used for the visualization of a time series visualization. A time series is a sequence of observations of the physical world in a regular time span. Time series help us to understand trends and correlations. As we can see in the following diagram, the vertical axis represents the value data and the horizontal axis represents time:

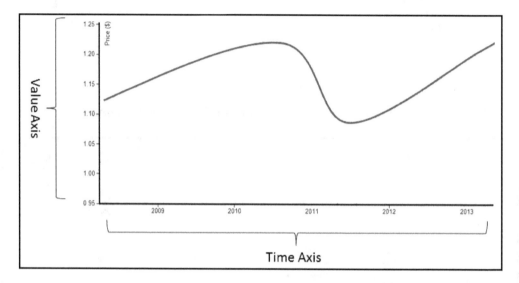

For this example, we will use the log of the USA/CAD historical exchange rates from March 2008 to March 2013 with 260 records.

In the following link, we can find the Historical Exchange Rates log to download:

http://www.oanda.com/currency/historical-rates/

The first seven records of the CSV file (line.csv) look like this:

```
date,usd
3/10/2013,1.0284
3/3/2013,1.0254
2/24/2013,1.014
2/17/2013,1.0035
2/10/2013,0.9979
2/3/2013,1.0023
```

```
1/27/2013,0.9973
    . . .
```

We need to define the font family and size for the labels and the style for the axis line:

```
<style>
body {
  font: 14px sans-serif;
}

.axis path,
.axis line {
  fill: gray;
  stroke: #000;
}
.line {
  fill: none;
  stroke: red;
  stroke-width: 3px;
}
</style>
```

Inside the body tag, we need to refer to the library:

```
<body>
<script src="http://d3js.org/d3.v3.js"></script>
```

We will define a format parser for the date value with d3.time.format. In this example, we have the data as follows: Month/Day/Year – "%m/%d/%Y" (e.g. 1/27/2013). Where %m represents the month as a decimal number from 01 to 12, %d represents the day of the month as a decimal number from 01 to 31, and %Y represents the year with century as a decimal number:

```
var formatDate = d3.time.format("%m/%d/%Y").parse;
```

To learn more about time formatting, please visit:
https://github.com/mbostock/d3/wiki/Time-Formatting/

Now we define the X and Y axes with a width of 1000 pixels and a height of 550 pixels:

```
var x = d3.time.scale()
    .range([0, 1000]);
var y = d3.scale.linear()
    .range([550, 0]);
```

```
var xAxis = d3.svg.axis()
    .scale(x)
    .orient("bottom");

var yAxis = d3.svg.axis()
    .scale(y)
    .orient("left");
```

The line element defines a line segment that starts at one point and ends at another.

> In the following link we can find the reference of SVG shapes:
> https://github.com/mbostock/d3/wiki/SVG-Shapes/

```
...var line = d3.svg.line()
    .x(function(d) { return x(d.date); })
    .y(function(d) { return y(d.usd); });
```

Now we select the element body and create a new element, `<svg>`, and define its size:

```
var svg = d3.select("body")
    .append("svg")
    .attr("width", 1000)
    .attr("height", 550)
  .append("g")
    .attr("transform", "translate("50,20")");
```

Then we need to open the file `"line.csv"` and read the values from the file and create the variable `data` with two attributes: `date` and `usd` as shown in the following snippet:

```
d3.csv("line.csv", function(error, data) {
data.forEach(function(d) {
   d.date = formatDate(d.date);
   d.usd = +d.usd;
 });
```

We define the date in the horizontal axis (`x.domain`) and in the vertical axis (`y.domain`) and set our value axis with the exchange rate value, `usd`:

```
x.domain(d3.extent(data, function(d) { return d.date; }));
y.domain(d3.extent(data, function(d) { return d.usd; }));
```

Finally, we add the groups of points and the labels in the axis:

```
svg.append("g")
    .attr("class", "x axis")
    .attr("transform", "translate(0,550)")
    .call(xAxis);
svg.append("g")
    .attr("class", "y axis")
    .call(yAxis)
svg.append("path")
    .datum(data)
    .attr("class", "line")
    .attr("d", line);
}); // close the block d3.csv
```

In the following screenshot we can see the result of the visualization:

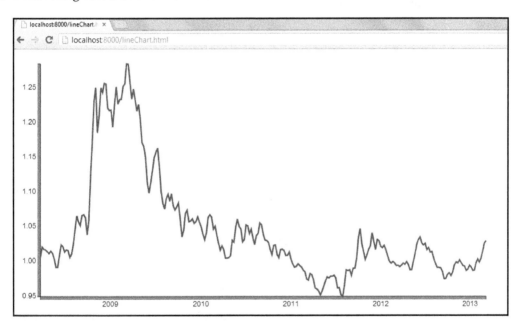

Multiple line chart

In a single variable we can see trends, but often we need to compare multiple variables and even find correlations or cluster trends. In this example, we will evolve the last example to work with multiple line charts. In this case, we will use data from historical exchange rates from the USA, Europe, and the UK.

In the following link we can find the historical exchange rates log ready for downloading:
http://www.oanda.com/currency/historical-rates/

The first five records of the CSV file (`multiline.csv`) look like this:

```
date,USD/CAD,USD/EUR,USD/GBP
03/10/2013,1.0284,0.7675,0.6651
03/03/2013,1.0254,0.763,0.6609
2/24/2013,1.014,0.7521,0.6512
2/17/2013,1.0035,0.7468,0.6402
02/10/2013,0.9979,0.7402,0.6361
. . .
```

We need to define the font family and size for the labels and the style for the axis line:

```
<style>
body {
   font: 18px sans-serif;
}
.axis path,
.axis line {
   fill: none;
   stroke: #000;
}
.line {
   fill: none;
   stroke-width: 3.5px;
}
</style>
```

Inside the `body` tag, we need to refer to the library:

```
<body>
<script src="http://d3js.org/d3.v3.js"></script>
```

We will define a format parser for the date value with `d3.time.format`. In this example, we have the data in this format: Month/Day/Year – `"%m/%d/%Y"` (for example, 1/27/2013):

```
var formatDate = d3.time.format("%m/%d/%Y").parse;
```

Now we define the X and Y axes with a width `1000` pixels and a height of `550` pixels:

```
var x = d3.time.scale()
    .range([0, 1000]);
var y = d3.scale.linear()
    .range([550, 0]);
```

We will define an array of color for each line:

```
var color = d3.scale.ordinal()
    .range(["#04B486", "#0033CC", "#CC3300"]);

var xAxis = d3.svg.axis()
    .scale(x)
    .orient("bottom");

var yAxis = d3.svg.axis()
    .scale(y)
    .orient("left");

var line = d3.svg.line()
    .interpolate("basis")
    .x(function(d) { return x(d.date); })
    .y(function(d) { return y(d.currency); });
```

Now we select the element body and create a new element, <svg>, and define its size:

```
var svg = d3.select("body")
    .append("svg")
    .attr("width", 1100)
    .attr("height", 550)
  .append("g")
    .attr("transform", "translate("50,20")");
```

Then, we need to open the file "multiLine.csv" and read the values from the file, and create the variable data with two attributes: date and color.domain as shown following:

```
d3.csv("multiLine.csv", function(error, data) {
  color.domain(d3.keys(data[0]).filter(function(key)
{return key !== "date"; }));
```

Now we apply the format function to the column date:

```
data.forEach(function(d) {
  d.date = formatDate(d.date);
});
```

Then we will define currencies as a separate array for each color line:

```
var currencies = color.domain().map(function(name) {
  return {
    name: name,
    values: data.map(function(d) {
      return {date: d.date, currency: +d[name]};
    })
```

```
        };
     });

     x.domain(d3.extent(data, function(d) { return d.date; }));
     y.domain
   ([d3.min(currencies, function(c) { return d3.min(c.values,
   function(v) { return v.currency; }); }),
      d3.max(currencies, function(c) { return d3.max(c.values, function(v) {
   return v.currency; }); })
     ]);
```

Now we add the groups of points, as well as the color and labels for each line as shown in the following snippet:

```
    svg.append("g")
       .attr("class", "x axis")
       .attr("transform", "translate(0,550)")
       .call(xAxis);
    svg.append("g")
       .attr("class", "y axis")
       .call(yAxis)
    var country = svg.selectAll(".country")
       .data(currencies)
     .enter().append("g")
       .style("fill", function(d) { return color(d.name); })
       .attr("class", "country");
```

Finally, we add a legend to the multiline series chart:

```
   country.append("path")
       .attr("class", "line")
       .attr("d", function(d) { return line(d.values); })
       .style("stroke", function(d) { return color(d.name); });
   country.append("text").datum(function(d)
   { return {name: d.name, value:  d.values[d.values.length - 1]}; })
       .attr("transform", function(d) {
   return "translate("+ x(d.value.date)+","+ y(d.value.currency)+")";
    })
       .attr("x", 10)
       .attr("y", 20)
       .attr("dy", ".50em")
       .text(function(d) { return d.name; });
   }); // close the block d3.csv
```

In the following screenshot we can see the result of the visualization:

Interaction and animation

D3 provides good support for interactions, transitions, and animations. In this example, we will focus on the basic way of adding transitions and interactions to our visualization. This time, we will use a very similar code to the bar chart example in order to demonstrate how easy it is to add interactivity to a visualization.

We need to define the font family, the size for the labels, and the style for the axis line:

```
<style>
body {
   font: 14px arial;
}
.axis path,
.axis line {
   fill: none;
   stroke: #000;
}
.bar {
   fill: gray;
}
</style>
```

Inside the body tag, we need to refer to the library:

```
<body>
<script src="http://d3js.org/d3.v3.min.js"></script>
var formato = d3.format("0.0");
```

Now we define the **X** and **Y** axes with a width of 1200 pixels and a height of 550 pixels:

```
var x = d3.scale.ordinal()
    .rangeRoundBands([0, 1200], .1);

var y = d3.scale.linear()
    .range([550, 0]);

var xAxis = d3.svg.axis()
    .scale(x)
    .orient("bottom");

var yAxis = d3.svg.axis()
    .scale(y)
    .orient("left")
    .tickFormat(formato);
```

Now we select the element body and create a new element, <svg>, and define its size:

```
var svg = d3.select("body").append("svg")
    .attr("width", 1200)
    .attr("height", 550)
  .append("g")
    .attr("transform", "translate(20,50)");
```

Then we need to open the file **TSV (Tab Separated Values)** "sumPokemons.tsv", read the values from the file and create the variable data with two attributes: type and amount as shown following:

```
d3.tsv("sumPokemons.tsv", function(error, data) {
  data.forEach(function(d) {
    d.amount = +d.amount;
  });
```

With the function map, we get our categorical values (type of Pokemon) for the horizontal axis (x.domain), and in the vertical axis (y.domain) we will set our value axis with the maximum value by type (in case there is a duplicate value):

```
x.domain(data.map(function(d) { return d.type; }));
y.domain([0, d3.max(data, function(d) { return d.amount; })]);
```

Now we will create an SVG group element, which is used to group SVG elements together with the tag, <g>. Then, we will use a transform function to define a new coordinate system for a set of SVG elements by applying a transformation to each coordinate specified in this set of SVG elements.

```
svg.append("g")
    .attr("class", "x axis")
    .attr("transform", "translate(0," + height + ")")
    .call(xAxis);

svg.append("g")
    .attr("class", "y axis")
    .call(yAxis)
```

Now we need to generate `.bar` elements and add them to SVG; then, with the `data(data)` function, for each value in the data we will call the `.enter()` function and add a `"rect"` element. D3 allows you to select groups of elements for manipulation through the `selectAll` function.

We want to highlight any bar just by clicking on it. First, we need to define a click with `.on('click', function).)`. Next, we will define the change of style for the bar highlighted with `.style('fill', 'red');`. In the following diagram we can see the highlighted bars Bug, Fire, Ghost, and Grass. Finally, we are going to set a simple animation using a transition, `transition().delay`, with a delay between the appearance of each bar. (See the bar chart on the following page.)

For a complete reference of selections, visit the following link:
https://github.com/mbostock/d3/wiki/Selections/

```
svg.selectAll(".bar")
    .data(data)
    .enter().append("rect")
     .on('click', function(d,i) {
    d3.select(this).style('fill','red');
})
    .attr("class", "bar")
    .attr("x", function(d) { return x(d.type); })
```

For a complete reference on transitions, visit the following link:
https://github.com/mbostock/d3/wiki/Transitions/

```
    .attr("width", x.rangeBand())
    .transition().delay(function (d,i){ return i * 300;})
    .duration(300)
    .attr("y", function(d) { return y(d.amount); })
    .attr("height", function(d) { return 550 - y(d.amount);})
    ;
}); // close the block d3.tsv
```

All the codes and datasets of this chapter may be found in the author's GitHub repository at:
https://github.com/hmcuesta/PDA_Book/tree/master/Chapter4

In the following diagram we can see the result of the visualization:

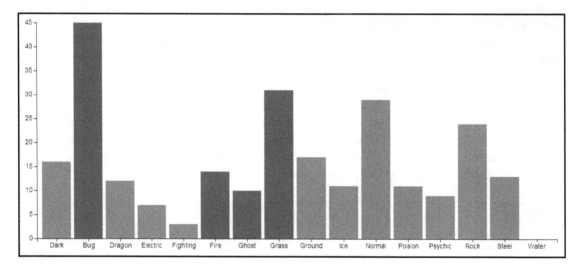

Data from social networks

Social networking sites represent a novel and valuable source of information. Increasing amounts of data on public events, points of interest, and urban sensing are available online. An unexpected explosion of user-generated content or information is available on social networks. An increase in the number of registered users increases the scope of industrial and academic research on social networks for different goals, including marketing, workload characterization, understanding and forecasting Internet use, and identifying the main challenges to support future Internet-based services.

One can gather a possible amount of information or content from social networks that may be useful for different situations, from marketing to intelligence. Unfortunately, the main challenge is collecting data or information from social networks with respect to other Internet-based sources. This might differ depending on the size and the complexity of a social network.

Extracting data from a social network requires an exploration of the user population with the purpose of collecting different kinds of information, such as comments, uploaded and downloaded content, ratings, and the network links among the users. Here, we outline some techniques proposed in literature for extracting data from social networks:

- Network traffic analysis [*Gill et al. (2007); Nazir et al. (2009)*]
- Ad-hoc applications [*Nazir et al. (2008)*]
- Crawling the user graph [*Mislove et al. (2007); Cha et al. (2008); Lerman (2007); Cqha et al. (2009)*]

An overview of visual analytics

Visual analytics can be seen as an integral approach combining visualization, data analysis, and human factors. In order to gain knowledge from data, the visual analytics procedure unites visual analysis methods and automatic processes through human interaction. In many application scenarios, visual or automatic analysis methods were applied after the integration of heterogeneous data sources. Therefore, before performing visual analysis we should clean, normalize, and integrate the heterogeneous data sources. After the data cleaning, the analyst may choose visual analysis methods, wherein visualization helps the analyst to relate with the automatic methods by modifying parameters or selecting other analysis algorithms. At the end, the model visualization can be used to study the findings of the generated models.

Summary

Visualization is an efficient way of finding frequent patterns or a relationship in a dataset. In this chapter, we have been introduced to a number of basic graphs implemented with D3.js. We discussed the most popular visualization techniques for discrete and continuous data. We explored the relationship between variables and saw how variables work over time. Similarly, we realized how to integrate basic user interaction and simple animation.

Finally, we discussed the concepts of lifting data from social networks and gave you an overview of visualization analytics and its uses, which will be covered in the ensuing chapters.

In the next chapter, you will be introduced to a variety of data analysis projects using machine learning algorithms as well as visualization tools.

4
Text Classification

This chapter builds on a brief introduction to text classification and provides you with an example of the **Naïve Bayes** algorithm, developed from scratch in order to explain how to turn an equation into code.

In this chapter, we will cover:

- Learning and classification
- Bayesian classification
- Naïve Bayes algorithm
- E-mail subject line tester
- The data
- The algorithm
- Classifier accuracy

Learning and classification

When we want to automatically identify which category belongs to a specific value (categorical value), we need to implement an algorithm that can decide the most likely category for the value based on previous data. This is called a **classifier**. In the words of Tom Mitchell:

> *"How can we build computer systems that automatically improve with experience, and what are the fundamental laws that govern all learning processes?"*

The key word here is learning (supervised learning, in this case) and knowing how to train an algorithm to identify categorical elements. The common examples are spam classification, speech recognition, search engines, computer vision, and language detection, but there is a large number of applications for a classifier. We can find two kinds of problems in classification. The **Binary classification** is where we only have two categories (**Spam** or **Not Spam**) and the **Multiclass classification**, in which there are many categories involved (such as the opinions Positive, Neutral, Negative, and so on). We can find several algorithms for classification; the most frequently used are support vector machines, neural networks, decision trees, NaÃ¯ve Bayes, and hidden Markov models. In this chapter, we will implement a probabilistic classification through the NaÃ¯ve Bayes algorithm, but in the following chapters, we will implement several other classification algorithms for a variety of problems.

The general steps involved in supervised classification are shown in the following diagram. First, we will collect training data (previously classified), then we will perform feature extraction (relevant features for the categorization), and next we will train the algorithm with the features vector. Once we get our trained classifier, we can insert new strings, extract its features, and send them to the classifier. Finally, the classifier will give us the most likely class (category) for the new string.

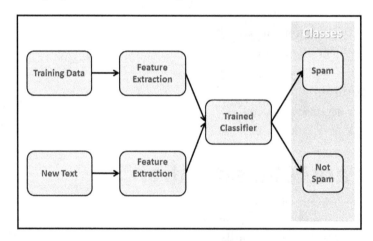

Additionally, we will test the classifier's accuracy by using a hand-classified test set. Due to this, we will split the data into two sets: the training data and the test data.

Bayesian classification

The probabilistic classification is a practical way to perform inferences based on data using statistical inferences to find the best class for a given value. Given a probability distribution, we can select the best option with the highest probability. The **Bayes Theorem** is the basic rule to perform inferences. The theorem allows us to update the likelihood of an event given the new data or observations. In other words, it allows us to update the prior probability $P (A)$ to the posterior probability $P (A|B)$. The prior probability is given by the likelihood before the data is evaluated and the posterior probability is assigned after the data is taken into account. The following expression represents the Bayes Theorem:

$$P(A \mid B) = \frac{P(B \mid A)P(A)}{P(B)}$$

$P(A|B) =$The conditional probability of **A** given **B**

Naïve Bayes

Naïve Bayes is the simplest classification algorithm among Bayesian classification methods. In this algorithm, we simply need to learn the probabilities by making the assumption that the attributes A and B are independents, hence the reason this model is defined as an *independent feature model*. Naïve Bayes is widely used in text classification because the algorithm can be trained easily and efficiently. In Naïve Bayes, we can calculate the probability of a condition A if B (P(A|B)), if you already know the probability of B given A (P(B|A)), as well as to A (P(A)) and B (P(B)) individually, as is shown in the previous Bayes Theorem example.

E-mail subject line tester

Spam is junk e-mail, which is understood as **Unsolicited Bulk Email** (**UBE**). E-mails can be blocked before they are delivered to the recipient based on e-mail filter reports. The e-mail filter scans the subject line of e-mails for spam or ham (e-mail that is not spam is often called ham). One of the e-mail filters is the e-mail subject line filter. Over 35 percent of spam mails are detected from the subject line of an e-mail.

An **E-mail Subject Line Tester** is a simple program that will define whether a certain subject line in an e-mail is spam or not. In this chapter, we will program a NaÃ¯ve Bayes classifier from scratch. The example will classify whether a subject line is a spam or not with a very simple code by breaking the subject lines into a list of relevant words that will be used as a feature vector in the algorithm. In order to do this, we will use the **SpamAssassin** public dataset. SpamAssasin includes three categories: **spam**, **easy ham**, and **hard ham**. In this case, we shall only use *spam* and *easy ham* (not spam) to train and test the classifier as we code a binary classifier.

There are several features that we can use for our classifiers such as the precedence, language, and the use of upper case. We will keep things simple and use the frequency of only words with more than three characters, and will avoid words such as *the* or *RT:* when training the algorithm.

We will implement the Bayes Rule using the words and categories, as shown in the following equation:

$$P(word \mid category) = \frac{P(category \mid word)P(word)}{P(category)}$$

For more information about probability distributions, please refer to:
`http://en.wikipedia.org/wiki/Probability_distribution`

We have two classes in the categories that represent whether a Subject Line is *Spam* or *Not Spam*. We need to split the texts into a list of words in order to get the likelihood of each word. Once we know the probability of each word, we need to multiply the probabilities for each category, as shown in the following equation:

$$P(category \mid word_1, word_2, ..., word_n) = P(category) x \Pi_i P(word \mid category)$$

In other words, we multiply the likelihood of each word *P(word|category)* of the subject line and the probability of the category *P(category)*.

For training the algorithm, we need to provide some prior examples. In this case, we use the function `training()`, which needs a dictionary of subject line and category, as we can see in the following examples:

```
Re: Tiny DNS Swap, nospam
Save up to 70% on international calls!, nospam
[Ximian Updates] Hyperlink handling in Gaim allows arbitrary code to be
executed, nospam
Promises., nospam
Life Insurance - Why Pay More?, spam
[ILUG] Guaranteed to lose 10-12 lbs in 30 days 10.206, spam
.  .  .
```

The data

We can find the spam dataset from the following link:

```
http://spamassassin.apache.org/
```

In the following screenshot, we can see the *easy ham* (not spam) folder with 2551 files:

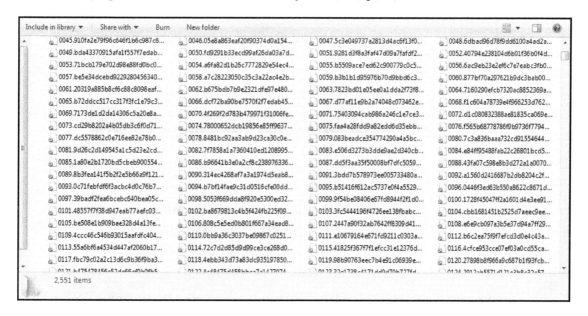

The spam text looks like the following screenshot, which may include HTML tags and plain text. In this case, we are only interested in the subject line, so we need to write a code to obtain the subject from all the files.

```
1    From smilee1313@eudoramail.com  Mon Aug 26 18:32:20 2002
2    Return-Path: <smilee1313@eudoramail.com>
3    Delivered-To: zzzz@localhost.spamassassin.taint.org
4    Received: from localhost (localhost [127.0.0.1])
5        by phobos.labs.spamassassin.taint.org (Postfix) with ESMTP id 4ABDD43F9B
6        for <zzzz@localhost>; Mon, 26 Aug 2002 13:32:20 -0400 (EDT)
7    Received: from mail.webnote.net [193.120.211.219]
8        by localhost with POP3 (fetchmail-5.9.0)
9        for zzzz@localhost (single-drop); Mon, 26 Aug 2002 18:32:20 +0100 (IST)
10   Received: from proxy-server.argogroupage.gr (mail.argogroupage.gr [195.97.102.134])
11       by webnote.net (8.9.3/8.9.3) with ESMTP id SAA27069
12       for <zzzz@spamassassin.taint.org>; Mon, 26 Aug 2002 18:30:18 +0100
13   Message-Id: <200208261730.SAA27069@webnote.net>
14   Received: from smtp0291.mail.yahoo.com (210.83.114.125 [210.83.114.125]) by proxy-s
15       id QP7CPKKZ; Sat, 24 Aug 2002 02:20:16 +0300
16   Date: Sat, 24 Aug 2002 07:08:34 +0800
17   From: "Jeannie Quiroz" <smilee1313@eudoramail.com>
18   X-Priority: 3
19   To: zzzz@netcomuk.co.uk
20   Cc: zzzz@spamassassin.taint.org, yyyy@netvision.net.il, yyyy@nevlle.net,
21       zzzz@news4.inlink.com
22   Subject: zzzz,Increase your breast size. 100% safe!
23   Mime-Version: 1.0
24   Content-Type: text/plain; charset=us-ascii
25   Content-Transfer-Encoding: 7bit
26
27   ===================================
28
29   Guaranteed to increase, lift and firm your
30   breasts in 60 days or your money back!!
31
32   100% herbal and natural.  Proven formula since
33   1996.  Increase  your bust by 1 to 3 sizes within 30-60
34   days and be all natural.
```

Hyper Text Markup Language file   length : 2118   lines : 57

This example will show you how to preprocess the SpamAssassin data using Python in order to collect all the subject lines from the e-mails.

First, we need to import the `os` module in order to get the list of file names using the `listdir` function from the " `\spam`" and " `\easy_ham`" folders:

```
import os
files = os.listdir(r" \spam")
```

Now we need a new file to store the subject lines and the category (spam or not spam); this time, we will use a comma as a separator:

```
with open("SubjectsSpam.out","a") as out:
    category = "spam"
```

Now we will parse each file and get the subject. Finally, we write the subject and the category in the new file and delete all the commas from the subject lines (`line.replace(",", "")`) to skip future troubles with the CSV format:

```
for fname in files:
    with open("\\spam" + fname) as f:
        data = f.readlines()
        for line in  data:
            if line.startswith("Subject:"):
                line.replace(",", "")
                print(line)
                out.write("{0}, {1} \n".format(line[8:-1], category))
```

We need to clean up the text, skipping the word *subject:* (the first eight characters) and the enter at the end of the text; in order to do this, we will use line[8:-1].

```
Output:
>>>Hosting from ?6.50 per month
>>>Want to go on a date?
>>>[ILUG] ilug,Bigger, Fuller Breasts Naturally In Just Weeks
>>> zzzz Increase your breast size. 100% safe!
. . .
```

We will keep the spam and not spam in different files to play with the size of the training and test sets. Usually, more data in the training set means better algorithm performance, but in this case, we will try to find an optimal threshold of the training set size.

All the codes and datasets of this chapter can be found in the author's GitHub repository at:
`https://github.com/hmcuesta/PDA_Book/tree/master/Chapter4`

The algorithm

We use the function `list_words()` to get a list of unique words with more than three characters in lower case:

```
def list_words(text):
    words = []
    words_tmp = text.lower().split()
    for w in words_tmp:
        if w not in words and len(w) > 3:
            words.append(w)
    return words
```

For a more advanced term-document matrix, we can use the Python textmining package from:
`https://pypi.python.org/pypi/textmining/1.`

The `training()` function creates variables to store the data needed for the classification. The `c_words` variable is a dictionary with the unique words and its number of occurrences in the text (frequency) by category. The `c_categories` variable stores a dictionary of each category and its number of texts. Finally, `c_text` and `c_total_words` store the total count of texts and words, respectively:

```
def training(texts):
    c_words ={}
    c_categories ={}
    c_texts = 0
    c_total_words =0
    #add the classes to the categories
    for t in texts:
        c_texts = c_texts + 1
        if t[1] not in c_categories:
            c_categories[t[1]] = 1
        else:
            c_categories[t[1]]= c_categories[t[1]] + 1

    #add the words with list_words() function
    for t in texts:
        words = list_words(t[0])

    for p in words:
        if p not in c_words:
            c_total_words = c_total_words +1
            c_words[p] = {}
            for c in c_categories:
                c_words[p][c] = 0
        c_words[p][t[1]] = c_words[p][t[1]] + 1

    return (c_words, c_categories, c_texts, c_total_words)
```

The `classifier()` function applies the Bayes rule and classifies the subject into one of the two categories, that is, either spam or not spam. The function also needs the four variables from the `training()` function:

```
def classifier(subject_line, c_words, c_categories, c_texts, c_tot_words):
    category =""
    category_prob = 0
```

```
for c in c_categories:
    #category probability
    prob_c = float(c_categories[c])/float(c_texts)
    words = list_words(subject_line)
    prob_total_c = prob_c
    for p in words:
        #word probability
        if p in c_words:
            prob_p= float(c_words[p][c])/float(c_tot_words)
            #probability P(category|word)
            prob_cond = prob_p/prob_c
            #probability P(word|category)
            prob =(prob_cond * prob_p)/ prob_c
            prob_total_c = prob_total_c * prob

    if category_prob < prob_total_c:
        category = c
        category_prob = prob_total_c
return (category, category_prob)
```

Finally, we will read the `training.csv` file, which contains the training dataset; in this case, 100 spam and 100 not spam subject lines:

```
if __name__ == "__main__":
    with open('training.csv') as f:
        subjects = dict(csv.reader(f, delimiter=','))
    words,categories,texts,total_words = training(subjects)
```

Now, to check if everything is working correctly, we test the classifier with one subject line:

```
clase = classifier("Low Cost Easy to Use Conferencing"
                   , words,categories,texts,total_words)

print("Result: {0} ".format(clase))
```

We can see the result in the Python console, and from the result we can see that the classifier is, so far, working correctly:

```
>>> Result: ('spam', 0.18518518518518517)
```

We can see the complete code of the NaÃ¯ve Bayes classifier listed here:

```
import csv
def list_words(text):
    words = []
    words_tmp = text.lower().split()
    for p in words_tmp:
        if p not in words and len(p) > 3:
```

```
            words.append(p)
    return words

def training(texts):
    c_words ={}
    c_categories ={}
    c_texts = 0
    c_tot_words =0
    for t in texts:
        c_texts = c_texts + 1
        if t[1] not in c_categories:
            c_categories[t[1]] = 1
        else:
            c_categories[t[1]]= c_categories[t[1]] + 1

    for t in texts:
        words = list_words(t[0])

    for p in words:
        if p not in c_words:
            c_tot_words = c_tot_words +1
            c_words[p] = {}
            for c in c_categories:
                c_words[p][c] = 0
        c_words[p][t[1]] = c_words[p][t[1]] + 1

    return (c_words, c_categories, c_texts, c_tot_words)


def classifier(subject_line, c_words, c_categories, c_texts, c_tot_words):
    category =""
    category_prob = 0

    for c in c_categories:
        prob_c = float(c_categories[c])/float(c_texts)
        words = list_words(subject_line)
        prob_total_c = prob_c
        for p in c_words:
            if p in words:
                prob_p= float(c_words[p][c])/float(c_tot_words)
                prob_cond = prob_p/prob_c
                prob =(prob_cond * prob_p)/ prob_c
                prob_total_c = prob_total_c * prob

            if category_prob < prob_total_c:
                category = c
                category_prob = prob_total_c
    return (category, category_prob)
```

```
if __name__ == "__main__":

    with open('training.csv') as f:
        subjects = dict(csv.reader(f, delimiter=','))

    w,c,t,tw = training(subjects)
    clase = classifier("Low Cost Easy to Use Conferencing"
                       ,w,c,t,tw)
    print("Result: {0} ".format(clase))
```

Finally, we will execute the classifier against the test set in order to prove accuracy. We will test 100 new records and get a new number of cases classified correctly.

```
with open("test.csv") as f:
    correct = 0
    tests = csv.reader(f)
    for subject in test:
        clase = classifier(subject[0],w,c,t,tw)
        if clase[1] =subject[1]:
        correct += 1
    print("Efficiency : {0} of 100".format(correct))
```

Classifier accuracy

Now we need to test our classifier with a bigger test set; in this case, we will randomly select 100 subjects: 50 spam and 50 not spam. Finally, we will count how many times the classifier chose the correct category:

```
with open("test.csv") as f:
    correct = 0
    tests = csv.reader(f)
    for subject in test:
        clase = classifier(subject[0],w,c,t,tw)
        if clase[1] =subject[1]:
        correct += 1
    print("Efficiency : {0} of 100".format(correct))
```

In this case, the `Efficiency` is 82 percent:

```
>>> Efficiency: 82 of 100
```

> We can use an out of the box implementation of the Naive Bayes classifier, like the `NaiveBayesClassifier` function in the NLTK package for Python. NLTK provides a very powerful natural language toolkit and we can download it from `http://nltk.org/`.

In Chapter 1, *Getting Started*, we presented a more sophisticated version of the NaÃ¯ve Bayes classifier to perform a sentiment analysis.

In this case, we will find an optimal size threshold for the training set. We will try a different number of random subject lines. In the following figure, we can see four of the seven tests and the classification rate (accuracy) of the algorithm. In all cases, we use the same test set of 100 elements and we use the same number of Email Subject Lines for each category:

- Test 1: 82 percent with a training set of 200 elements
- Test 2: 85 percent with a training set of 300 elements
- Test 5: 87 percent with a training set of 500 elements
- Test 7: 92 percent with a training set of 800 elements

We run several trials with a different number of records to train the classifier, and we can see that the accuracy is improved with a bigger training set:

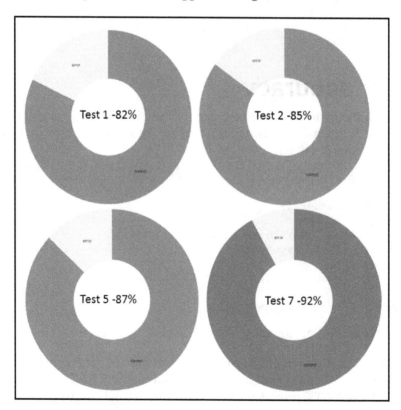

As we can see in the following diagram, for this specific example, the maximum accuracy is **92** percent and the optimal number of texts in the training set is 700. After 700 texts in the training set, the accuracy of the classifier doesn't see a significant improvement:

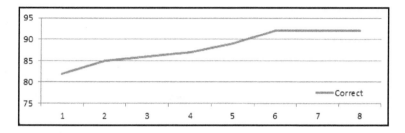

Summary

In this chapter, we created a basic but useful e-mail subject line tester. This chapter provides a guide on how to code a basic NaÃ¯ve Bayes classifier from scratch, without any external library, in order to demonstrate how easy it is to program a machine learning algorithm. We also defined a maximum size threshold for the training set and got an accuracy of 92 percent, which for this basic example is quite good.

In the following chapters, we will introduce more complex machine learning algorithms using the mlpy library and we will present how to extract more sophisticated features.

5
Similarity-Based Image Retrieval

A big part of the data that we work with is presented as an image, a drawing, or a photo. In this chapter, we will implement a **Similarity-Based Image Retrieval** without the use of any metadata or concept-based image indexing. We will work with distance metric and dynamic warping to retrieve the most similar images.

In this chapter, we will cover:

- Image similarity search
- Dynamic time warping
- Processing the image dataset
- Implementing DTW
- Analyzing the results
- Summary

Image similarity search

With the emergence of the Internet of Things and autonomous robots, the ability to compare an image from the environment to others from our database is fundamental for understanding its context. Imagine an autonomous car or a drone with a camera, looking for something like a red bag or an advertisement. This kind of task requires an image search without any kind of metadata associated. Due to this, you don't look for equality; instead, you look for similarity.

While comparing two or more images, the first question that comes to our mind is what makes an image similar to another? We can say that one image is equal to another if all their pixels match. However, a small change in the light, angle, or rotation of the camera represents a big change in the numerical values of the pixels. Finding ways to define whether two images are similar is the main concern of services such as Google Search by Image or TinEye, where the user uploads an image instead of providing keywords or descriptions as the search criteria.

Humans have natural mechanisms to detect patterns and similarity. Comparing images at content or semantic levels is a difficult problem and an active research field in computer vision, image processing, and pattern recognition. We can represent an image as a matrix (two-dimensional array), in which each position of the matrix represents the intensity or the color of the image. However, any change in the lighting, camera angle, or rotation means a large numerical shift in the matrix. The question that comes to mind is how can we measure similarity between matrices? To address these problems, the data analysis implements several **Content-Based Image Retrieval (CBIR)** tools, such as comparison of wavelets, Fourier analysis, or pattern recognition with neural networks. However, these methods cause a loss of a lot of the image information or require extensive training, like with neural networks. The most commonly used method is **Description-Based Image Retrieval (DBIR)**, which uses metadata associated with the images, but in an unknown dataset, this method is not effective.

In this chapter, we used a different approach, taking advantage of the elastic matching of a time series, which is a method widely used in voice recognition and time series comparison. For the purpose of this chapter, we will understand time series as a sequence of pixels. The trick is to turn the pixels of the image into a numerical sequence, as shown in the following image displayed:

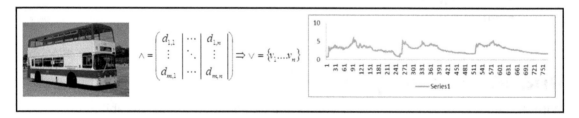

Dynamic time warping

Dynamic Time Warping (**DTW**) is an elastic matching algorithm used in pattern recognition. DTW finds the optimal warp path between two vectors. DTW is used as a distance metric often implemented in speech recognition, data mining, robotics, and in this case, image similarity. The distance metric measures how far from each other two points, **A** and **B**, are in a geometric space. We commonly use the Euclidian Distance, which draws a direct line between a pair of points. In the following image, we might see different kinds of paths between point **A** and **B**, such as the Euclidian distance (with the arrow), but we also see the Manhattan (or taxicab) distance (with the dotted lines). Imagine that you are looking for the shortest path to a destination—the straight line will not work due to all the buildings, so the Taxicab distance simulates the way a New York taxi navigates through the buildings, as shown in the following diagram:

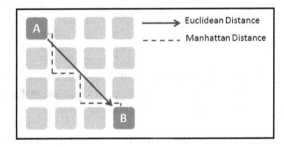

DTW is used to define the similarity between time series for classification; in this example, we will implement the same metric with sequences of pixels. We can say that if the distance between sequence A and sequence B is small, these images are similar. We will use the Manhattan distance between the two series to sum off the squared distances. However, we can use other distance metrics such as Minkowski or Euclidean (that choice depends on the problem at hand).

> Distance metrics are formulated in the Taxicab geometry proposed by Hermann Minkowski; for more information about it, visit:
> `http://taxicabgeometry.net/`

In the following diagram, we can observe a warping between two vectors:

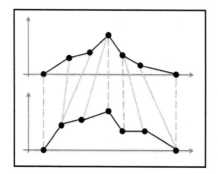

The example of this chapter will use **mlpy**, which is a Python module for machine learning built on top of **NumPy** and **SciPy**. The **mlpy** library implements a version of DTW that can be found at:

`http://mlpy.sourceforge.net/docs/3.4/dtw.html`

In the paper *Direct Image Matching by Dynamic Warping*, Hansheng Lei and Venu Govindaraju implement a DTW for image matching finding an optimal pixel to pixel alignment and prove that DTW is very successful in the task.

can observe a cost matrix with the minimum distance warp path traced through it to indicate the optimal alignment:

In the following screenshot, we can observe a cost matrix with the minimum distance warp path traced through it to indicate the optimal alignment:

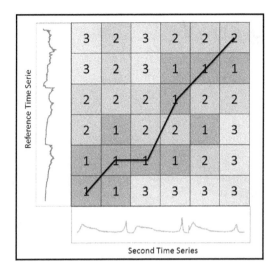

Processing the image dataset

The image set used in this chapter is the **Caltech-256** obtained from the **Computational Vision Lab** at CALTECH. We can download the collection of all 30607 images and 256 categories from the following link:

```
http://www.vision.caltech.edu/Image_Datasets/Caltech256/
```

In order to implement the DTW first, we need to extract a time series (pixel sequences) from each image. The time series will have a length of 768 values and will add the 256 values of each color in the **RGB** (Red, Green, and Blue) color model of each image. The following code implements the `Image.open("Image.jpg")` function and casts it into an array, and then simply adds the three vectors of color in the list:

```
from PIL import Image
img = Image.open("Image.jpg")
arr = array(img)
list = []
for n in arr: list.append(n[0][0])  #R
for n in arr: list.append(n[0][1])  #G
for n in arr: list.append(n[0][2])  #B
```

Pillow is a PIL fork by Alex Clark and is compatible with Python 2.x and 3.x. **PIL** is the **Python Imaging Library** by Fredrik Lundh. In this chapter, we will use Pillow due to its compatibility with Python 3.2, and it can be downloaded from `https://github.com/python-imaging/Pillow`.

Implementing DTW

In this example, we will look for a similarity in 684 images from eight categories. We will use four imports of `PIL`, `numpy`, `mlpy`, and `collections`:

```
from PIL import Image
from numpy import array
import mlpy
from collections import OrderedDict
```

First, we need to obtain the time series representation of the images and store it in a dictionary (`data`) with the number of the image and its time series `data[fn] = list`:

```
data = {}
for fn in range(1,685):
    img = Image.open("ImgFolder\\{0}.jpg".format(fn))
    arr = array(img)
    list = []
    for n in arr: list.append(n[0][0])
    for n in arr: list.append(n[0][1])
    for n in arr: list.append(n[0][2])
    data[fn] = list
```

The performance of this process will lie in the number of images processed, so beware of the use of memory with large datasets.

Then, we need to select an image for reference, which will be compared to all the other images in the dictionary `data`:

```
reference = data[31]
```

Now we need to apply the `mlpy.dtw_std` function to all the elements and store the distance in the result dictionary:

```
result ={}
for x, y in data.items():
    #print("{0} --------------- {1}".format(x,y))
    dist = mlpy.dtw_std(reference, y, dist_only=True)
    result[x] = dist
```

Finally, we need to sort the result in order to find the closest elements using the function `OrderedDict` and then we can print the ordered result:

```
sortedRes = OrderedDict(sorted(result.items(), key=lambda x: x[1]))
for a,b in sortedRes.items():
    print("{0}-{1}".format(a,b))
```

In the following screenshot, we can see the result and we can observe that the result is accurate with the first element (reference time series). The first result presents a distance of `0.0` because it's exactly the same as the image we used as a reference.

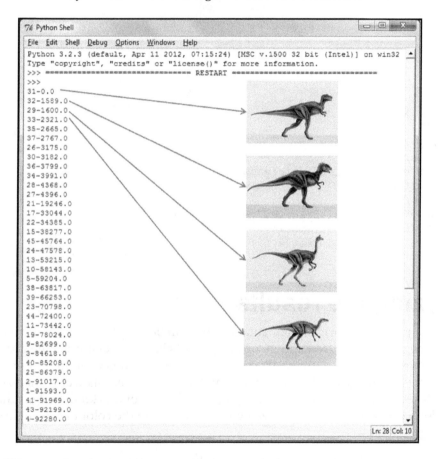

All the codes and datasets of this chapter can be found in the author's GitHub repository at:
https://github.com/hmcuesta/PDA_Book/tree/master/Chapter5

We can see the complete code here:

```
from PIL import Image
from numpy import array
import mlpy
from collections import OrderedDict

data = {}
for fn in range(1,685):
    img = Image.open("ImgFolder\\{0}.jpg".format(fn))
    arr = array(img)
    list = []
    for n in arr: list.append(n[0][0])
    for n in arr: list.append(n[0][1])
    for n in arr: list.append(n[0][2])
    data[fn] = list
reference = data[31]

result ={}

for x, y in data.items():
    #print("{0} --------------- {1}".format(x,y))
    dist = mlpy.dtw_std(reference, y, dist_only=True)
    result[x] = dist

sortedRes = OrderedDict(sorted(result.items(), key=lambda x: x[1]))
for a,b in sortedRes.items():
    print("{0}-{1}".format(a,b))
```

Analyzing the results

This example presents a basic implementation that can be adapted in several cases, such as 3D object recognition, face recognition, or as a part of clustering analysis. The goal of this chapter is to present how we can easily compare vectors unsupervised in order to find the similarity between images. In this section, we will present seven cases and analyze the results. This result will help us understand the possibilities of this kind of algorithm, for example, for a drone to find a target even with variations in the color or the angle of the picture.

In the following screenshot, we can see the first three searches and we can observe high accuracy in the result; even in the case of a bus, the result displays the result elements in different angles, rotations, and colors:

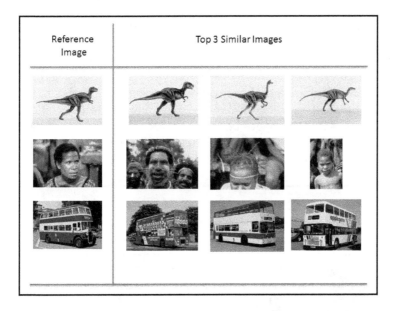

In the following screenshot, we see the searches 4 (horse), 5 (flower), and 6 (elephant), and we can observe that in an image with good contrast in colors, the algorithm performs well:

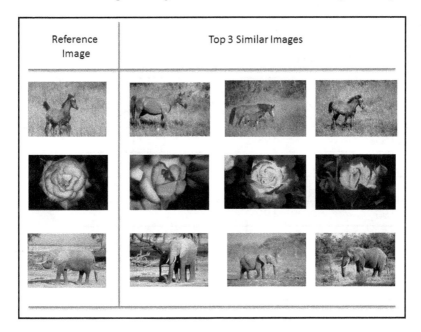

In case of the search 7 (the building), the result is poor, and in similar cases, when the reference's time series is a landscape or a building, the result is often images that are not related to the search criteria. This is because the RGB color model of the time series is very similar to other categories. In the following screenshot, we can observe that the reference image and the first result share a high saturation of the color blue; due to this, their time series (sequences of pixels) are very similar. We may overcome this problem by using a filter, such as *Find Edges*, in the images before the search. In `Chapter 14`, *Online Data Analysis with Jupyter and Wakari,* we will present the use of filters, operations, and transformations for image processing using PIL.

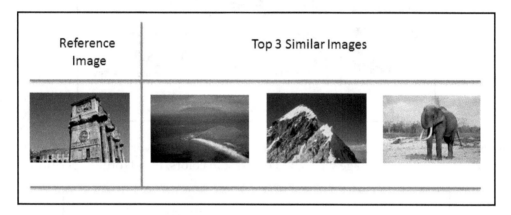

In the following table, we can see the results of the complete set of tests:

Categories	Number of Images	% of First Result Right	% of Second Result Right
Dinosaurs	102	99	99
African people	85	98	95
Bus	56	98	90
Horse	122	92	88
Roses	95	96	92
Elephants	36	98	87
Landscape	116	60	52
Buildings	72	50	45

Summary

In this chapter, we introduced the **Dynamic Time Warping (DTW)** algorithm, which is an excellent tool for finding similarity between vectors with an unsupervised learning model. We presented an implementation of DTW to find similarity between a set of images, which works very well in most cases. This method can be used in several other problems, in a variety of areas, such as robotics, computer vision, speech recognition, and time series analysis. We also saw how to turn an image into a time series with the PIL library. Finally, we learned how to implement DTW with the Python mlpy library.

In the next chapter, we will present how simulation can help us in data analysis and how to model pseudo-random events.

6
Simulation of Stock Prices

The author *Burton Malkiel* proposed in his top-seller book *A Random Walk Down Wall Street (1973)* that a stock price will take a random path and that we cannot use historical data to predict stock future, due to its behavior being independent of other factors. The **Random Walk** theory may help us to simulate this kind of unpredictable path. In this chapter, we will implement a simulation of stock prices applying the Random Walk algorithm and implement it with a D3.js animation. This chapter will cover the following topics:

- Financial time series
- Random Walk simulation
- Monte Carlo methods
- Generating random numbers
- Implementation in D3.js
- Quantitative Analysts

Financial time series

Financial Time Series Analysis (FTSA) involves working with asset valuation over time, such as currency exchange or stock market prices. FTSA addresses a particular feature: uncertainty. In the words of famous American financier J. P. Morgan, when asked what the stock market will do, he said.

> *"It will fluctuate"*

The uncertainty of financial time series means that the volatility of a stock price cannot be directly observable. In fact, *Louis Bachelier's Theory of Speculation* (1900) postulated that prices fluctuate randomly.

In the following diagram, we can see the **Apple Inc.** time series of historical stock prices for three months. In fact, simple random processes can create a time series that will closely resemble this real-time series. The Random Walk model is considered in FTSA as a statistical model for the movement of logged stock prices.

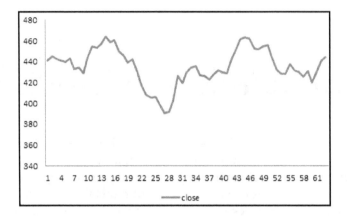

We can download Apple Inc.'s historical stock prices from the Nasdaq website at `http://www.nasdaq.com/symbol/aapl/historical#.UT1jrRy pwJ`.

For a complete and in-depth reference, see the book *Analysis of Financial Time Series* by *Ruey S. Tsay*. In this chapter, we will implement a random walk simulation in D3.js, and in the following section we will discuss random walk and Monte Carlo models.

Random Walk simulation

Random Walk is a simulation where a succession of random steps are used to represent an apparently random event. The interesting thing is that we can use this kind of simulation to see different outputs from a certain event by controlling the start point of the simulation and the probability distribution of the random steps. Like all simulations, this is just a simplified model of the original phenomena. However, a simulation might be useful and is a powerful visualization tool. There are different notions of Random Walks using different implementations, with the most common being **Brownian motion** and the **Binomial model**. We will use these models to visualize the random path followed by stock prices through time.

In the following diagram, we can see simulated data from the Random Walk model for logged stock prices:

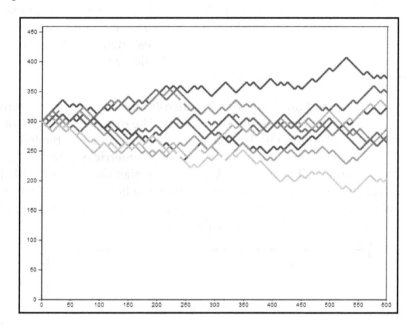

Brownian motion is a Random Walk model named after physicist Robert Brown, who observed molecules moving and colliding with one another in a random fashion. Brownian motion is usually used to model stock prices. According to the work of Robert C. Merton (Nobel laureate in Economics), the **Brownian model of financial markets** defines that stock prices evolve continuously in time and are driven by Brownian motion processes. In this model, a normal distribution of a returns period is assumed; this means that the probability of the random-step does not vary over time and is independent of past steps.

Binomial Model is a simple price model that is based on discrete steps where the prices of an asset can go up or down. If the prices go up, then it is multiplied by an up-factor, and on the other hand, if the asset goes down, then it is multiplied by a down-factor.

For more information on the Brownian model of financial markets, please refer to `http://bit.ly/17WeyH7`.

Monte Carlo methods

Random Walk is a member of a family of random sampling algorithms. Proposed by *Stanislaw Ulam* in 1940, **Monte Carlo** methods are mainly used when the event has uncertainty and deterministic boundaries (the previous estimate was for a range of limit values). These methods are especially good for optimization and numerical integration in finance, biology, business, physics, and statistics.

Monte Carlo methods depend on the probability distribution of the random number generator to see different behaviors in the simulations. The most common distribution is the **Gauss** or **Normal**; this distribution is also referred to as *Bell Curve* (see the following diagram), but there are more distributions such as the **Geometric** or **Poisson**. In statistics, the **Central Limit Theorem** (CTL) proposes that the Gaussian distribution will appear in almost any case. Where the sample of n elements from a uniform random source (if the number of samples gets larger, the approximation improves), the sum of these values will be distributed accordingly to a Gaussian distribution.

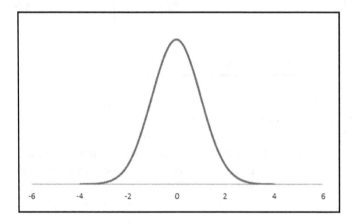

Generating random numbers

While getting truly random numbers is a difficult task, most of the Monte Carlo methods perform well with pseudo-random numbers, and this makes it easier to rerun simulations based in a seed. Practically, all modern programming languages include basic random sequences, or at least sequences good enough to produce accurate simulations.

- Python includes the `random` library. In the following code, we can see the basic usage of this library:

```
import random as rnd
```

- Getting a random float between 0 and 1:

```
>>>rnd.random()
0.254587458742659
```

- Getting a random number between 1 and 100:

```
>>>rnd.randint(1,100)
56
```

- Getting a random float between 10 and 100 using a uniform distribution:

```
>>>rnd.uniform(10,100)
15.2542689537156
```

> For a detailed list of methods in the random library, go to: `http://docs.python.org/3.2/library/random.html`

In the case of JavaScript, a more basic random function is included with the function `Math.random()`; for the purpose of this chapter, it will be enough.

In the following script, we can see a basic JavaScript code printing a random number between 0 and 100 in an HTML element with the ID `"label"`:

```
<script>
function  randFunction()
{
var x=document.getElementById("label")
x.innerHTML=Math.floor((Math.random()*100)+1);
}
</script>
```

Implementation in D3js

In this chapter, we will create an animation in D3.js of a Brownian motion Random Walk simulation. In the simulation, we will control the delay of the animation, the starting point of the Random Walk, and the tendency of the up-down factor.

First, we need to create an HTML file named `Simulation.html`, and we will run it from Python `http.server`. In order to run the animation, we just need to open a Command Terminal and run the following command:

```
>>python -m http.server 8000
```

Open a web browser and type the direction `http://localhost:8000` and select our HTML file; then, we can see the animation running.

First, we need to import the D3 library either directly from the website or with a local copy of the `d3.v3.min.js` file:

```
<script type="text/javascript" src="http://d3js.org/d3.v3.min.js"></script>
```

In the CSS, we specified the style for the axis line, the font family and size for the text, and the background color:

```
<style type="text/css">
body {
  background: #fff;
}
.axis text {
  font: 10px sans-serif;
}
.axis path,
.axis line {
  fill: none;
  stroke: #000;
}
</style>
```

 We can define the colors in CSS using a hexadecimal code like `#fff` instead of the literal name *white*; we can also find a color selector at: `http://www.w3schools.com/tags/ref_colorpicker.asp`

We need to define some variables that we will need in the animation, such as the delay, the first line color, and the height and width of the work area. The color variable will be randomly reassigned every time the time series reaches the edge of the canvas, so we start with a new color for the next time series:

```
var color = "rgb(0,76,153)";
var GRID = 6,
HEIGHT = 600,
WIDTH = 600,
delay = 50;
```

Now we need to define the size of the new SVG width and height (630 x 650 pixels, including extra space for the axis labels); this inserts a new `<svg>` element inside the `<body>` tag:

```
var svg = d3.select("body").append("svg:svg")
   .attr("width", WIDTH + 50)
   .attr("height", HEIGHT + 30)
  .append("g")
   .attr("transform", "translate(30,0)");
```

Now we need to set the associated scale for the X and Y axes as well as the labels orientation:

```
var x = d3.scale.identity()
    .domain([0, WIDTH]);
var y = d3.scale.linear()
    .domain([0, HEIGHT])
    .range([HEIGHT, 0]);
var xAxis = d3.svg.axis()
    .scale(x)
    .orient("bottom")
    .tickSize(2, -HEIGHT);
var yAxis = d3.svg.axis()
    .scale(y)
    .orient("left")
    .tickSize(6, -WIDTH);
```

Apply the axis to an SVG selection with the element `<g>`:

```
svg.append("g")
    .attr("class", "x axis")
    .attr("transform", "translate(0,600)")
    .call(xAxis);
    svg.append("g")
    .attr("class", "y axis")
    .attr("transform", "translate(0,0)")
    .call(yAxis);
```

> We can define the D3 reference API documentation of SVG axes at `https://github.com/mbostock/d3/wiki/SVG-Axes`.

Then we will include a text label in the **X** axis (position 270, 50):

```
svg.append("text")
        .attr("x", 270 )
        .attr("y",  50 )
        .style("text-anchor", "middle")
        .text("Random Walk Simulation");
```

We will create a function called `randomWalk` to perform each step of the simulation. This will be a recursive function and will include the drawing of the line segments for each step of the Random Walk. Using the `Math.random()` function, we will decide whether the walker goes up or down:

```
function randomWalk(x, y) {
var x_end, y_end = y + GRID;
if (Math.random() < 0.5) {
  x_end = x + GRID;
} else {
  x_end = x - GRID;
}
line = svg.select('line[x1="' + x + '"][x2="' + x_end + '"]'+
                  '[y1="' + y + '"][y2="' + y_end + '"]');
```

Now we need to add the new line segment to the SVG element `"svg:line"`, with a random color and 3 points of stroke width:

```
svg.append("svg:line")
    .attr("x1", y)
    .attr("y1", x)
    .attr("x2", y_end)
    .attr("y2", x_end)
    .style("stroke", color)
    .style("stroke-width", 3)
    .datum(0);
```

When the walker (`y_end`) reaches the end of the work space, we need to pick a new color randomly with the function `Math.floor(Math.random()*254)` in each of the RGB codes and then reset the control variables (`y_end` and `x_end`):

```
if (y_end >= HEIGHT) {
color = "rgb("+Math.floor(Math.random()*254)+",
            "+Math.floor(Math.random()*254)+",
            "+Math.floor(Math.random()*254)+")"
  x_end = WIDTH / 2;
  y_end = 0;
}
```

With the `window.setTimeout` function, we will wait 50 milliseconds to get the progressive effect of the animation and call the `randomWalk` function again:

```
window.setTimeout(function() {
   randomWalk(x_end, y_end);
}, delay);
}
```

Finally, we need to call the `randomWalk()` function, passing as a parameter the starting point in the Y axis of the animation:

```
randomWalk(WIDTH / 2, 0);
```

> All the code in this chapter can be found in the author's GitHub repository at:
> `https://github.com/hmcuesta/PDA_Book/tree/master/Chapter6`

In the following screenshot, we can see the result of the animation after twelve iterations in **1** and after much iteration in **2**:

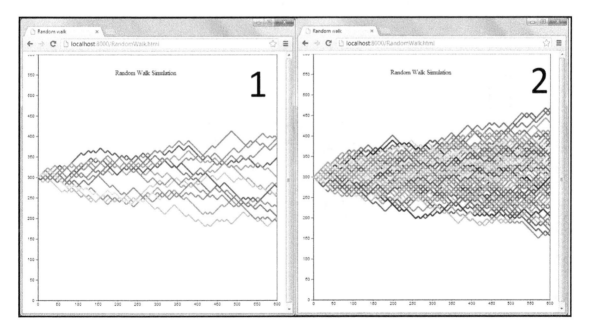

One interesting thing that we can observe is that the Gaussian distribution is presented in the visualization. In the following diagram, we can see the normal distribution of the Random Walk in a shaded area:

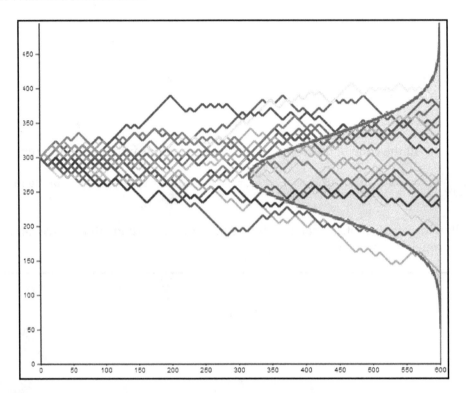

We can also try with different start parameters to get different outputs, such as by changing the start point of the lines and the distribution of the Random Walk, as we can see in the following screenshot:

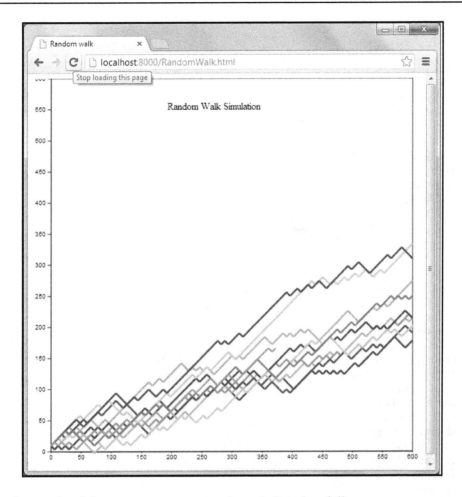

The complete code of the `Random Walk` simulator is listed as follows:

```html
<html>
  <head>
    <meta content="text/html;charset=utf-8">
    <title>Random walk</title>
    <script type="text/javascript"    src="http://d3js.org/d3.v3.min.js">
</script>
    <style type="text/css">
 body {
  background: #fff;
}
.axis text {
  font: 10px sans-serif;
}
```

```
.axis path,
.axis line {
  fill: none;
  stroke: #000;
}
      </style>
</head>
<body>
<script>
var color = "rgb(0,76,153)";
var GRID = 6,
HEIGHT = 600,
WIDTH = 600,
delay = 50,
svg = d3.select("body").append("svg:svg")
  .attr("width", WIDTH + 50)
  .attr("height", HEIGHT + 30)
 .append("g")
  .attr("transform", "translate(30,0)");
var x = d3.scale.identity()
    .domain([0, WIDTH]);
var y = d3.scale.linear()
    .domain([0, HEIGHT])
    .range([HEIGHT, 0]);
var xAxis = d3.svg.axis()
    .scale(x)
    .orient("bottom")
    .tickSize(2, -HEIGHT);
var yAxis = d3.svg.axis()
    .scale(y)
    .orient("left")
    .tickSize(6, -WIDTH);
svg.append("g")
    .attr("class", "y axis")
    .attr("transform", "translate(0,0)")
    .call(yAxis);
svg.append("g")
    .attr("class", "x axis")
    .attr("transform", "translate(0,600)")
    .call(xAxis);
svg.append("text")
        .attr("x", 270 )
        .attr("y",  50 )
        .style("text-anchor", "middle")
        .text("Random Walk Simulation");
    randomWalk(WIDTH / 2, 0);
    function randomWalk (x, y) {
    var x_end, y_end = y + GRID;
```

```
if (Math.random() < 0.5) {
  x_end = x + GRID;
} else {
  x_end = x - GRID;
}  line = svg.select('line[x1="' + x + '"][x2="' + x_end + '"]'+
                '[y1="' + y + '"][y2="' + y_end + '"]');
  svg.append("svg:line")
    .attr("x1", y)
    .attr("y1", x)
    .attr("x2", y_end)
    .attr("y2", x_end)
    .style("stroke", color)
    .style("stroke-width", 3)
    .datum(0);
  if (y_end >= HEIGHT) {
  color = "rgb("+Math.floor(Math.random()*254)+",
            "+Math.floor(Math.random()*254)+",
            "+Math.floor(Math.random()*254)+")"
  x_end = WIDTH / 2;
  y_end = 0;
}
  window.setTimeout(function() {
    randomWalk(x_end, y_end);
  }, delay);
}
</script>
</body>
</html>
```

Quantitative analyst

Quantitative finance is an amazing area that applies mathematical models to financial markets; example tasks include algorithmic trading, derivatives pricing, and risk management. This requires people with financial, computer, and analytical skills; this person is called a quantitative analyst or a quant. A quantitative analyst is in charge of proposing testable hypotheses and develops models to uncover hidden patterns. Quantitative finance helps to apply the scientific method to the study of financial markets using simulations like those we used in this chapter, or predictive methods like the ones we will use in Chapter 7, *Predicting Gold Prices*.

Summary

In this chapter, we explored the Random Walk simulation and how to communicate through animated visualizations. Simulation is an excellent way to observe the behavior of a phenomenon such as stock prices. The Monte Carlo methods are widely used to simulate phenomena when we don't have the means to reproduce an event because it is dangerous or expensive, like with epidemic outbreaks or stock prices. However, a simulation is always a simplified model of the real world. The goal of the simulation presented in this chapter is to show how we can get a basic but attractive web-based visualization with D3.

In the next chapter, we will learn the basic concepts of time series. Then, we will present a numeric prediction of gold prices using regression and classification techniques.

7
Predicting Gold Prices

In this chapter, you will be introduced to the basic concepts of **Time Series Data** and **Regression**. First, we distinguish some of the basic concepts such as **Trend**, **Seasonality**, and **Noise**, along with the principles of **Lineal Regression** using the Python library **scikit-learn**. Then, we will introduce the **Historic Gold Prices** time series and see how to perform a nonlinear forecast using **Kernel Ridge Regression**. Later, we will present a regression using the smoothed time series as the input.

This chapter will cover the following topics:

- Working with time series data
- Lineal regression
- The data: historical gold prices
- Nonlinear regression
- Kernel Ridge Regression
- Smoothing the gold prices time series
- Predicting in the smoothed time series
- Contrasting the predicted value

Working with time series data

Time series is one of the most common ways to find data in the real world. A time series is defined as the changes of a variable through time. **Time Series Analysis** (**TSA**) is widely used in economics, finance, weather, and epidemiology. If we look at one TSA, we may see different kinds of patterns and discover outliers easily. Working with time series needs to define some basic concepts of trend, seasonality, and noise.

In the following graph, we can see the time series for gold price in the US since July 2010 from `http://www.gold.org/investment/statistics/gold_price_chart/`.

Typically, the easiest way to explore a time series is with a line chart. With the help of the direct appreciation of the time series visualization, we can find anomalies and complex behavior in the data:

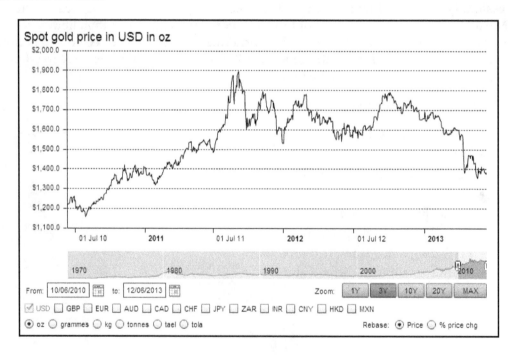

We have two kinds of time series: **linear** and **nonlinear**. In the following graph, we can see an example of each one. Plotting time series data is very similar to scatterplot or line chart, but the data points in axis **X** are either times or dates:

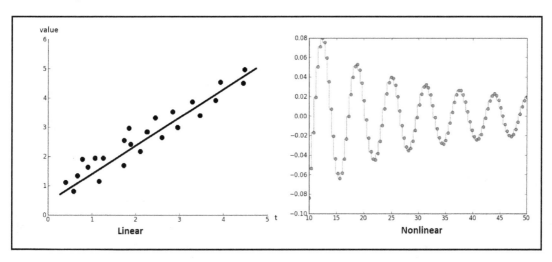

Components of a time series

In many cases, the time series is the sum of multiple components:

$$X_t = T_t + S_t + V_t$$

Here, *Observation = Trend + Seasonality + Variability* where:

- **Trend (T)** is the behavior or slow motion in the time series through a large timeframe
- **Seasonality (S)** is the oscillatory motion in a year, for example, the flu season
- **Variability (V)** is the random variations around the previous components

In the following graph, we can see a time series with an evolutionary trend that doesn't follow a linear pattern and slowly evolves through time:

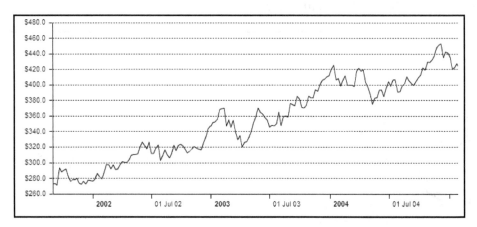

In this book, the visualization is driven with D3.js (web based); however, it is important to have a fast visualization tool directly from the Python language. In this chapter, we will use `matplotlib` as a standalone visualization tool. In the following code, we can see an example of how to use `matplotlib` to visualize a line chart.

First, we need to import the library and assign an alias `plt` library:

```
import matplotlib.pyplot as plt
```

Then, using the `numpy` library, we will create a synthetic data with the `linspace` and `cos` methods for the x and y data respectively:

```
import numpy as np
x = np.linspace(10, 100, 500)
y = np.cos(x)/x
```

Now, we will prepare the visualization with the `step` function; and with the `show` function, we will present the visualization in a new window:

```
plt.step(x, y)
plt.show()
```

 We can find more information about `matplotlib` at `http://matplotlib.org/`

Finally, we can see the visualization window with the result in the following screenshot:

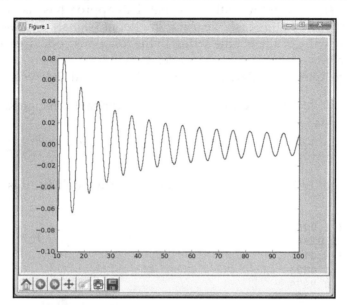

As we can see in the preceding screenshot, the visualization window provides us with some tools such as pan axes, zoom, and save, which help us to prepare and export the visualization in a `.png` image format. We can also navigate through the changes or go back to the original view. If you are using an interactive environment, such as IPython Notebook, you should use the `%matplotlib` inline instruction in order to see the graph in the same window.

Smoothing time series

When we work with some real-world data, we might often find noise that is defined as pseudo random fluctuations in values that don't belong to the observation data. In order to avoid or reduce this noise, we can use different approaches, such as increasing the amount of data by the interpolation of new values, where the series is sparse; however, in many cases, this is not an option. Another approach is smoothing the series, typically using the `averages` or `exponential` method. The `average` method helps us smooth the series by replacing each element in the series with either the simple, or the weighted average of the data around it. We will define the **Smoothing Window** to the interval of possible values, which controls the smoothness of the result. The main disadvantage of using the moving averages approach is that if we have outliers or abrupt jumps in the original time series, the result might be inaccurate and can produce jagged curves.

In this chapter, we will implement a different approach using convolution (moving averages filter) of a scaled window with the signal. This approach is taken from **Digital Signal Processing** (**DSP**). In this case, we use a time series (signal), and we will apply a filter; as a result, we will get a new time series. In the following code, we can see an example of how to smooth a time series. For this example, we will use the log of USA/CAD Historical Exchange Rates from March 2008 to March 2013, with 260 records.

 In the following link, we can find the Historical Exchange Rates log to be downloaded:
http://www.oanda.com/currency/historical-rates/

The first seven records of the CSV file (`ExchangeRate.csv`) look like this:

```
date,usd
3/10/2013,1.028
3/3/2013,1.0254
2/24/2013,1.014
2/17/2013,1.0035
2/10/2013,0.9979
2/3/2013,1.0023
1/27/2013,0.9973
    .  .  .
```

First, we need to import all the required libraries; see `Appendix`, *Setting Up the Infrastructure*, for the complete installation instructions on the `numpy` and `scipy` libraries:

```
import dateutil.parser as dparser
import matplotlib.pyplot as plt
import numpy as np
from pylab import *
```

Now, we will create the `smooth` function that will send as parameters the original time series and the window's length. In this implementation, we use the `numpy` implementation of the `hamming` window (`np.hamming`); however, we can use other kinds of windows such as **flat**, **hanning**, **Bartlett**, and **blackman**.

 For a complete reference of the supported window functions by Numpy, refer to http://docs.scipy.org/doc/numpy/reference/routines.wind ow.html.

Take a look at following code snippet:

```
def smooth(x,window_len):
        s=np.r_[2*x[0]-x[window_len-1::-1],
            x,2*x[-1]-x[-1:-window_len:-1]]
        w = np.hamming(window_len)
        y=np.convolve(w/w.sum(),s,mode='same')
        return y[window_len:-window_len+1]
```

The method presented in this chapter is based on the signal smoothing from the scipy reference documentation that can be found at http://wik i.scipy.org/Cookbook/SignalSmooth.

Then, we need to obtain the labels for the **X** axis, using the numpy genfromtxt function to get the first column in the csv file. We need to apply a converter function called dparser.parse to parse the date data:

```
x = np.genfromtxt("ExchangeRate.csv",
                dtype='object',
                delimiter=',',
                skip_header=1,
                usecols=(0),
                converters = {0: dparser.parse})
```

Now, we need to obtain the original time series from the "ExchangeRate.csv" file:

```
originalTS = np.genfromtxt("ExchangeRate.csv",
                skip_header=1,
                dtype=None,
                delimiter=',',
                usecols=(1))
```

Then, we apply the smooth method and store the result in the smoothedTS list:

```
smoothedTS = smooth(originalTS, len(originalTS))
```

Finally, we plot the two series using pyplot:

```
plt.step(x, originalTS, 'co')
plt.step(x, smoothedTS)
plt.show()
```

In the followinggraph, we can see the original (dotted line) and the smoothed (line) series. We can observe in the visualization that in the smoothed series, we cut out the irregular roughness to see a clearer signal. Smoothing doesn't provide us with a model per se. However, it can be the first step to describe multiple components of the time series. When we work with epidemiological data, we can smooth out the seasonality so that we can identify the trend (See `Chapter 10`, *Working with Social Graphs*):

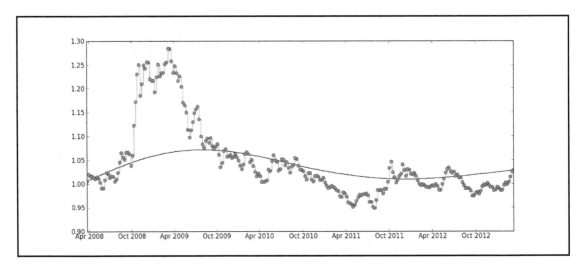

Lineal regression

If we want to predict a quantitative value, **regression** is a great tool due to it uses. It's an independent variable to explain the behavior of a phenomenon such as temperature, asset prices, house prices, and so on. Linear regression finds the best fitting in a straight line.

We use regression or forecast all the time in our daily lives: when we calculate the gas or the time required for a car trip based on previous data (distance, traffic, weather, and so on). In its simplest form, you can think of it in this way: first, get previous data from the phenomena, for example, how much time was spent on previous trips and what was the distance. Then, look at the values form, and try to find a metric to forecast the next value.

In this section, we will program a very simple example of linear regression using **scikit-learn**, which is a machine-learning library for Python. For this concrete example, we will use the **Boston Housing dataset**, which represents the data of 506 neighborhoods of Boston, and is included in the scikit-learn library or is available at `https://archive.ics.uci.edu/ml/datasets/Housing`.

First, we need to import all the required libraries: the `pylab`, `linear_model`, and `boston` datasets:

```
from pylab import * from sklearn import datasets
from sklearn import linear_model
from sklearn.cross_validation import train_test_split
import numpy as np
 import matplotlib.pyplot as plt
```

Check out the following link for a complete reference, and download the instructions for scikit-learn:
`http://scikit-learn.org/stable/`

Then, we need to acquire the data from the scikit-learn datasets, and we select just one independent feature to perform the regression, such as price, land size, or number of rooms:

```
houses = datasets.load_boston()
houses_X = houses.data[:, np.newaxis]
houses_X_temp = houses_X[:, :, 2]
```

Now, we need to split the dataset into a training set and a test subset with a test size of 33 percent of the total amount of the rows. The training set will help us teach the algorithm and the patterns in the data for the forecast:

```
X_train, X_test, y_train, y_test = train_test_split(houses_X_temp,
houses.target, test_size=0.33)
```

Then, we select the algorithm for the `LinearRegression()` forecast, and then we train the algorithm with the training set using the `fit()` function:

```
lreg = linear_model.LinearRegression()
lreg.fit(X_train, y_train)
```

Finally, we will plot the result. We might see a straight line, which is the predicted data (blue) for the test set (green). This is a very simple way to perform a regression due to the data being restricted to only one independent variable. In the real world, the phenomenon is modeled with several variables using multidimensional algorithms:

```
plt.scatter(X_test, y_test, color='green')
plt.plot(X_test, lreg.predict(X_test), color='blue', linewidth=2)
plt.show()
```

Take a look at the following screenshot:

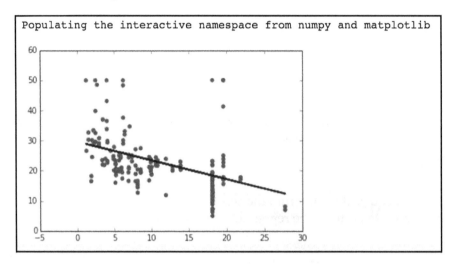

The data – historical gold prices

Regression analysis is a statistical tool to understand the relationship between variables. In this chapter, we will implement a **nonlinear regression** to predict the gold price based on historical gold prices. For this example, we will use the historical gold prices from January 2003 to May 2013 in a monthly range obtained from www.gold.org. Finally, we will forecast the gold price for June 2013, and we will contrast it with the real price from an independent source. The complete dataset (since December 1978) is at http://www.gold.org/research/ download-the-gold-price-since-1978.

The first seven records of the CSV file (gold.csv) look like this:

```
date,price
1/31/2003,367.5
2/28/2003,347.5
3/31/2003,334.9
4/30/2003,336.8
5/30/2003,361.4
6/30/2003,346.0
7/31/2003,354.8
    . . .
```

In this example, we will implement a Kernel Ridge Regression, which is a regression method based on nonlinear kernels. We will use the original time series and the smoothed time series to compare the differences in the output.

Nonlinear regressions

Statistically speaking, the nonlinear regression is a kind of regression analysis used to estimate the relationships between one or more independent variables in a nonlinear combination.

In this chapter, we will use the `mlpy` Python library, and its Kernel Ridge Regression implementation. We can find more information about nonlinear regression methods at `http://mlpy.sourceforge.net/docs/3.3/nonlin_regr.html`.

Kernel Ridge Regressions

The most basic algorithm that can be kernelized is (**KRR**Kernel Ridge Regression (KRR), which is a combination of **Ridge Regression** using a small kernel trick that corresponds to a nonlinear function that fits a line to some values mapped from **X** to **Y**. It is similar to a **Support Vector Machines** (**SVM**), as we will see in `Chapter 8`, *Working with Support Vector Machines*, but the solution depends on all the training samples and not on a subset of support vectors. KRR works well with a few training sets for classification and regression. It is widely used for recommendation systems, face recognition, and regression models. Due to this, it gives great results when you're trying to model data that has a small number of values or nonlinear data points. In this chapter, we will focus on its implementation using `mlpy` rather than all the linear algebra involved. See `Appendix`, *Setting Up the Infrastructure*, for complete installation instructions on the `mlpy` library.

First, we need to import the `numpy`, `mlpy`, and `matplotlib` libraries:

```
import numpy as np
import mlpy
from mlpy import KernelRidge
import matplotlib.pyplot as plt
```

Now, we define the seed for the random number generation:

```
np.random.seed(10)
```

Then, we need to load the historical gold prices from the "`Gold.csv`" file and store them in `targetValues`:

```
targetValues = np.genfromtxt("Gold.csv",
                    skip_header=1,
                    dtype=None,
                    delimiter=',',
                    usecols=(1))
```

Next, we will create a new array with `125` training points—one for each record of `targetValues`, representing the monthly gold price from January 2003 to May 2013:

```
trainingPoints = np.arange(125).reshape(-1, 1)
```

Then, we will create another array with `126` test points representing the original `125` points in `targetValues`, and including an extra point for our predicted value for June 2013:

```
testPoints = np.arange(126).reshape(-1, 1)
```

Now, we create the training kernel matrix (`kn1`) and testing kernel matrix (`kn1Test`). KRR will randomly split the data into subsets of the same size. Then, it will process an independent KRR estimator for each subset. Finally, it averages the local solutions into a global predictor:

```
kn1 = mlpy.kernel_gaussian(trainingPoints, trainingPoints,
                    sigma=1)
kn1Test = mlpy.kernel_gaussian(testPoints, trainingPoints,
                    sigma=1)
```

Then, we instantiate the `mlpy.KernelRidge` class in the `kn1Ridge` object:

```
kn1Ridge = KernelRidge(lmb=0.01, kernel=None)
```

The `learn` method will compute the regression coefficients using the training kernel matrix and the target values as a parameter:

```
kn1Ridge.learn(kn1, targetValues)
```

The `pred()` method computes the predicted response using the testing kernel matrix as an input:

```
resultPoints = kn1Ridge.pred(kn1Test)
```

Finally, we plot the two time series of target values and result points:

```
fig = plt.figure(1)
plot1 = plt.plot(trainingPoints, targetValues, 'o')
plot2 = plt.plot(testPoints, resultPoints)
plt.show()
```

In the following graph, we can see that the points represent the target values (the known values) and the line represents the result points (result from the `pred` method). We can also see the last segment of the line, which is the predicted value for June 2013:

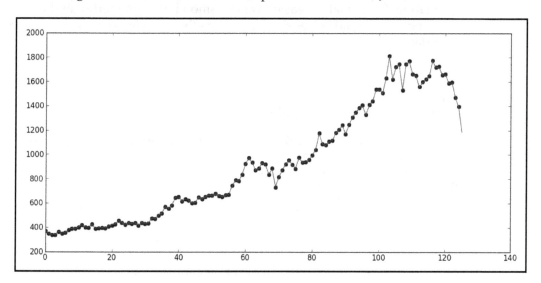

In the following screenshot, we can see the resulting points from the `knlRidge.pred()` method, and the last value (`1186.16129538`) is the predicted value for June 2013:

```
74 "Python Shell"                                                    □ ⊡ ✕
File  Edit  Shell  Debug  Options  Windows  Help
     992.90306715   1043.00379077   1170.80325892   1089.47935716   1076.97972427
    1107.14368599   1116.03303904   1176.36610067   1208.7325989    1239.51413576
    1171.27046153   1242.32094619   1307.12838129   1343.38595492   1383.85097939
    1400.21052961   1329.35732296   1405.38790921   1440.13464059   1530.38444854
    1535.76768304   1501.3482692    1629.64604751   1803.83015783   1622.14643238
    1717.76197169   1739.11892089   1534.37590801   1736.64798768   1767.27956776
    1660.76742321   1646.36358188   1557.34365693   1595.2421219    1618.05827517
    1648.59106968   1768.32721459   1720.07173547   1718.51077847   1658.84824093
    1657.43062521   1590.26281453   1591.23005356   1470.40013319   1389.83693915
    1186.16129538]
                                                                    Ln: 31  Col: 0
```

All the code and datasets of this chapter can be found in the author's GitHub repository at `https://github.com/hmcuesta/PDA_Book/tree/master/Chapter7`.

Smoothing the gold prices time series

As we can see, the gold prices time series is noisy, and it's hard to spot trends or patterns with a direct appreciation. So, to make it easier, we can smooth the time series. In the following code, we smooth the gold prices time series (see the *Smoothing time series* section in this chapter for a detailed explanation):

```python
import matplotlib.pyplot as plt
import numpy as np
import dateutil.parser as dparser
from pylab import *
def smooth(x,window_len):
        s=np.r_[2*x[0]-x[window_len-1::-1],x,2*x[-1]-x[-1:-window_len:-1]]
        w = np.hamming(window_len)
        y=np.convolve(w/w.sum(),s,mode='same')
        return y[window_len:-window_len+1]
x = np.genfromtxt("Gold.csv",
                    dtype='object',
                    delimiter=',',
                    skip_header=1,
                    usecols=(0),
                    converters = {0: dparser.parse})
y = np.genfromtxt("Gold.csv",
                    skip_header=1,
                    dtype=None,
                    delimiter=',',
                    usecols=(1))
y2 = smooth(y, len(y))
plt.step(x, y2)
plt.step(x, y, 'co')
plt.show()
```

In the following graph, we can see the time series of the historical gold prices (the dotted line), and we can see the smoothed time series (the line) using the hamming window:

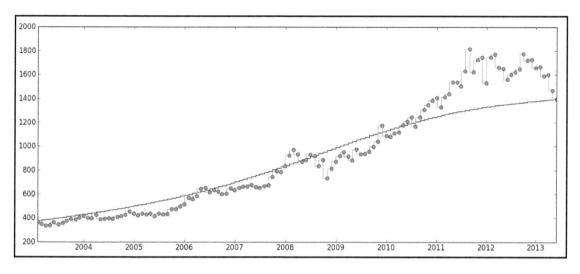

Predicting in the smoothed time series

Finally, we will put them all together and implement the KRR to the smoothed gold prices time series. We can find the complete code of the KRR here:

```
import matplotlib.pyplot as plt
import numpy as np
import dateutil.parser as dparser
from pylab import *
import mlpy
def smooth(x,window_len):
        s=np.r_[2*x[0]-x[window_len-1::-1],
x,2*x[-1]-x[-1:-window_len:-1]]
        w = np.hamming(window_len)
        y=np.convolve(w/w.sum(),s,mode='same')
        return y[window_len:-window_len+1]
y = np.genfromtxt("Gold.csv",
                    skip_header=1,
                    dtype=None,
                    delimiter=',',
                    usecols=(1))
targetValues = smooth(y, len(y))
np.random.seed(10)
trainingPoints = np.arange(125).reshape(-1, 1)
testPoints = np.arange(126).reshape(-1, 1)
```

```
knl = mlpy.kernel_gaussian(trainingPoints,
                           trainingPoints, sigma=1)
knlTest = mlpy.kernel_gaussian(testPoints,
                               trainingPoints, sigma=1)
knlRidge = mlpy.KernelRidge(lmb=0.01, kernel=None)
knlRidge.learn(knl, targetValues)
resultPoints = knlRidge.pred(knlTest)

plt.step(trainingPoints, targetValues, 'o')
plt.step(testPoints, resultPoints)
plt.show()
```

In the following graph, we can see the dotted line that represents the smoothed time series of the historical gold prices, and the line that represents the prediction for the gold price in June 2013:

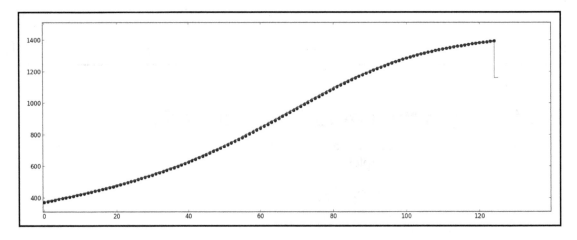

In the following screenshot, we can see the predicted values for the smoothed time series. This time, we can observe that the values are much lower than the original predictions:

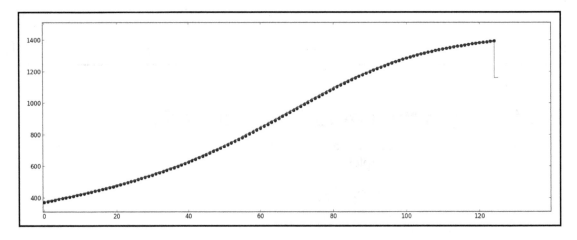

Contrasting the predicted value

Finally, we will look for an external source to see if our prediction is realistic. In the following graph, we can observe a graph from The Guardian/Thomson Reuters for June 2013, and see that the gold price fluctuated between 1180.0 and 1210.0, with an official average in the month of 1192.0. Our prediction for the KRR with complete data is 1186.0, which is not bad at all. We can see the complete numbers in the following table:

....	June 2013
Guardian/Thomson Reuters	1192.0
Kernel Ridge Regression with Complete Data	1186.161295
Kernel Ridge Regression with Smoothed Data	1159.23545044

For this particular example, the predicted value using the complete data is more accurate than that predicted with the smoothed data. When working with machine learning models, trying different approaches is very important in order to find the most accurate algorithm for a specific dataset.

In the words of the mathematician George E. P. Box:

"All models are wrong, but some are useful"

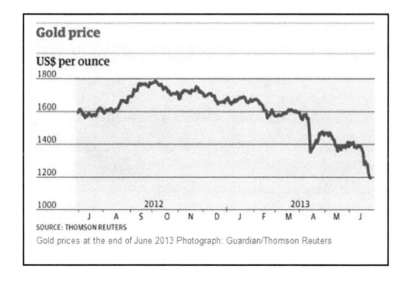

For complete information about the *Stock markets and gold suffer a June to forget* article, refer to `http://www.theguardian.com/business/213/jun/28/stock-markets-gold-june`.

Summary

In this chapter, we explored the nature of time series, described their components, and implemented signal processing to smooth the time series. We examined a simple example of linear regression using scikit-learn. Then, we introduced the KRR implemented in the `mlpy` library. Finally, we presented two implementations of KRR, one with with the complete data and the other with the smoothed data, to predict the monthly gold price in June 2013. We found that, for this case, the prediction with the complete data was more accurate.

In the next chapter, you will learn how to perform a dimensionality reduction and how to implement a SVM with a multivariate dataset.

8
Working with Support Vector Machines

The **Support Vector Machine** (**SVM**) is a powerful classification technique based on Kernels, such as the **Kernel Ridge Regression** (KRR) algorithm seen in the previous chapter. We often deal with sparse datasets or with data that is not good enough to make a good prediction or classification. In such cases, we may use a technique that creates new values from the original dataset to help in the accuracy of the algorithm; this new data is called synthetic. Due to their efficiency, using Kernels is one of the most common methods to make synthetic data. In this chapter, we will provide you with an easy way to get acceptable results using SVM. We will perform a dimensionality reduction of the dataset, and we will produce a model for classification.

The theoretical foundation of SVM lies in the work of Vladimir Vapnik and the theory of statistical learning developed in the 70s. SVMs are highly used in pattern recognition of Time Series, Bioinformatics, Natural Language Processing, and Computer Vision.

In this chapter, we will use the `mlpy` implementation of **LIBSVM**, which is a widely used library for SVM with several interfaces and extensions for languages, such as Java, Python, Matlab, R, CUDA, C#, and Weka. For more information about LIBSVM, you can visit `http://www.csie.ntu.edu.tw/~cjlin/libsvm/`.

In this chapter, we will cover the following topics:

- Understanding the multivariate dataset
- Dimensionality reduction
- Getting started with Support Vector Machines
- Kernel functions

- Double spiral problem
- SVM implementation using mlpy

> Another great implementation for SVM is included in the **scikit-learn** Python library. Follow the link to see the full reference:
> `http://scikit-learn.org/stable/modules/svm.html`

Understanding the multivariate dataset

A multivariate dataset is defined as a set of multiple observations (attributes) associated with different aspects of a phenomenon. In this chapter, we will use a multivariate dataset, which is the result of a chemical analysis of wines that grew in three different cultivars from the same area in Italy. The **Wine** dataset is available in the **UC Irvine Machine Learning Repository** and can be freely downloaded from `http://archive.ics.uci.edu/ml/datase ts/Wine`. This dataset includes physicochemical data from white and red wine from the north of Portugal in order to find quality levels. The dataset includes 13 features with no missing data, and all the features are numerical or real values.

The complete list of features is listed here:

- Alcohol
- Malic acid
- Ash
- Alkalinity of ash
- Magnesium
- Total phenols
- Flavanoids
- Nonflavanoid phenols
- Proanthocyanins
- Color intensity
- Hue
- OD280/OD315 of diluted wines
- Proline

The classes in the dataset are ordered and not balanced; this means that the number of rows is not in the same proportion. The dataset has 178 records from three different classes. The distribution is seen in the following diagram, corresponding to 59 for class 1, 71 for class 2, and 48 for class 3:

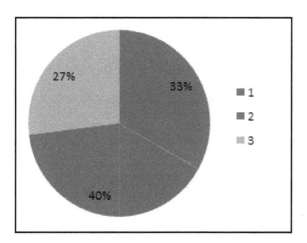

The first five records of the dataset will look like this:

```
1,14.23,1.71,2.43,15.6,127,2.8,3.06,.28,2.29,5.64,1.04,3.92,1065
1,13.2,1.78,2.14,11.2,100,2.65,2.76,.26,1.28,4.38,1.05,3.4,1050
1,13.16,2.36,2.67,18.6,101,2.8,3.24,.3,2.81,5.68,1.03,3.17,1185
1,14.37,1.95,2.5,16.8,113,3.85,3.49,.24,2.18,7.8,.86,3.45,1480
1,13.24,2.59,2.87,21,118,2.8,2.69,.39,1.82,4.32,1.04,2.93,735
```

In the following code, we will plot two of the features from the dataset at a time. In this example, we will plot the `Alcohol` and `Malic acid` attributes. However, to visualize all the possible feature combinations, we will need the binomial coefficient of the number of features. In this case, 13 features are equal to 78 different combinations. Due to this, it is mandatory to perform a dimensionality reduction:

```
import matplotlib
import matplotlib.pyplot as plt
```

First, we will import the data from the dataset into a matrix of the features and a list of the categories associated with each record using the `getData` function:

```
def getData():
    lists = [line.strip().split(",") for line in open('wine.data',
'r').readlines()]
    return [list( l[1:14]) for l in lists], [l[0] for l in lists]
matrix, labels = getData()
```

```
xaxis1 = []; yaxis1 = []
xaxis2 = []; yaxis2 = []
xaxis3 = []; yaxis3 = []
```

Then, we will select the two features to visualize in the x and y variables:

```
x = 0 #Alcohol
y = 1 #Malic Acid
```

Next, we will generate the sets of coordinates for the two attributes (x,y) of the three categories (classes) of the dataset:

```
for n, elem in enumerate(matrix):
    if int(labels[n]) == 1:
        xaxis1.append(matrix[n][x])
        yaxis1.append(matrix[n][y])
    elif int(labels[n]) == 2:
        xaxis2.append(matrix[n][x])
        yaxis2.append(matrix[n][y])
    elif int(labels[n]) == 3:
        xaxis3.append(matrix[n][x])
        yaxis3.append(matrix[n][y])
```

Finally, we will plot the three categories (classes) into a scatter plot:

```
fig = plt.figure()
ax = fig.add_subplot(111)
type1 = ax.scatter(xaxis1, yaxis1, s=50, c='white')
type2 = ax.scatter(xaxis2, yaxis2, s=50, c='red')
type3 = ax.scatter(xaxis3, yaxis3, s=50, c='darkred')
ax.set_title('Wine Features', fontsize=14)
ax.set_xlabel('X axis')
ax.set_ylabel('Y axis')
ax.legend([type1, type2, type3], ["Class 1", "Class 2", "Class 3"], loc=1)
ax.grid(True,linestyle='-',color='0.80')
plt.show()
```

In the following diagram, we can observe the plot result of the features `Alcohol` and `Malic Acid` for the three classes:

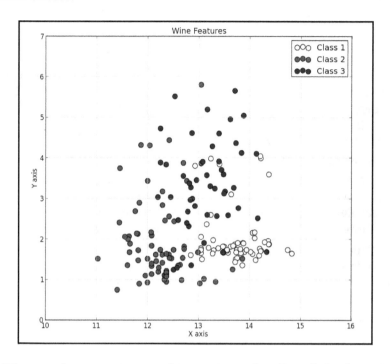

We may also use a scatter plot matrix to visualize all the features in the dataset. However, this is computationally expensive and the number of subplots will depend on the binomial coefficient of the number of features. See `Chapter 14`, *Online Data Analysis with Jupyter and Wakari*, to see how to plot multidimensional datasets with a scatterplot matrix and **RadViz**.

Dimensionality reduction

The dimensionality of a model is the number of independent attributes in the dataset. In order to reduce the complexity of the model, we need to reduce the dimensionality without sacrificing accuracy. When we work in complex multidimensional data, we need to select the features that can improve the accuracy of the technique we are using. Sometimes, we don't know whether the variables are independent or if they share some kind of relationship. We need criteria to find the best features and reduce the number of variables under consideration.

In order to address these problems, we will perform three techniques: **Feature selection**, **Feature extraction**, and **Dimension reduction**:

- **Feature selection**: We will select a subset of features in order to get better training times or improve the model accuracy. In data analysis, finding the best features for our problem is often guided by intuition, and we don't know the real value of a variable until we test it. However, we may use metrics such as **correlation** and **mutual information**, which can help us by providing distance between features. The correlation coefficient is a measure of how strong the relationship is between two variables and mutual information refers to a measure of how much one variable tells about another.

- **Feature extraction**: This is a special form of dimensionality reduction performed by a transformation in a high-dimensional space (multivariate dataset), to get a space of fewer dimensions (the ones that are more informative). Two of the classical algorithms in the field are **Principal Component Analysis (PCA)** or **Multidimensional scaling (MDS)**. Feature extraction is widely used in image processing, computer vision, and data mining.

- **Dimension reduction**: When we work with high-dimensional data, there are various phenomena that may affect the result of our analysis; this is known as the curse of dimensionality. In order to avoid these problems, we will apply a preprocessing step using **Principal Component Analysis (PCA)** or **Linear Discriminant Analysis (LDA)**.

We can find more information about the curse of dimensionality at `http:/ /bit.ly/7xJNzm`.

Linear Discriminant Analysis (LDA)

LDA is a statistical method used to find a linear combination of features, which can be used as a linear classifier. LDA is often used as a dimensionality reduction step before a complex classification. The main difference between LDA and PCA is that PCA does feature extraction and LDA performs classification. The `mlpy` implementation of LDA can be found at `http://bit.ly/19xyq3H`.

Principal Component Analysis (PCA)

PCA is the most used dimensionality reduction algorithm. PCA is an algorithm used to find a subset of features, linearly uncorrelated, known as principal components. In order to make the dataset understandable in an easy way, we can reduce the number of variables to two interpretable linear combinations of the data. Each linear combination will correspond to a principal component. PCA can be used in **exploratory data analysis** (EDA), through visual methods, to find the most important characteristics in a dataset. This time, we will implement a feature selection and PCA to the Wine dataset. In the following code, we present the basic implementation of PCA in `mlpy`.

First, we need to import the `numpy`, `mlpy`, and `matplotlib` modules as shown in the following code:

```
import numpy as np
import mlpy
import matplotlib.pyplot as plt
import matplotlib.cm as cm
```

Next, we will open `wine.data` using the `numpy` function `loadtxt`:

```
wine = np.loadtxt('wine.data', delimiter=',')
```

Then, we will define the features; in this case, we will select the features 2 (`Malic acid`), 3 (`Ash`), and 4 (`Alkalinity of ash`) for axis **X**, and the class (which is the feature) for axis **Y**:

```
x, y = wine[:, 2:5], wine[:, 0].astype(np.int)
print(x.shape)
print(y.shape)
```

In this case, `x.shape` and `y.shape` will look like this:

```
>>> (178,3)
>>> (178,)
```

Now, we will create a new instance of `PCA`, and we need to train the algorithm with selected features in x using the learn function:

```
pca = mlpy.PCA()
pca.learn(x)
```

Then, we will apply the dimensionality reduction to the features in the x variable, and we turn it into two-dimensional subspace with the `k = 2` parameter:

```
z = pca.transform(x, k=2)
```

The result of the transformation will be stored in the z variable, and its shape will look like this:

```
print(z.shape)
>>> (178,2)
```

Finally, we will use matplotlib to visualize the scatter plot of the new two-dimensional subspace of PCA stored in z:

```
fig1 = plt.figure(1)
title = plt.title("PCA on wine dataset")
plot = plt.scatter(z[:, 0], z[:, 1], c=y, s=90, cmap=cm.Reds)
labx = plt.xlabel("First component")
laby = plt.ylabel("Second component")
plt.show()
```

In the following diagram, we can observe the scatter plot of the PCA result using green, blue, and red to highlight each class:

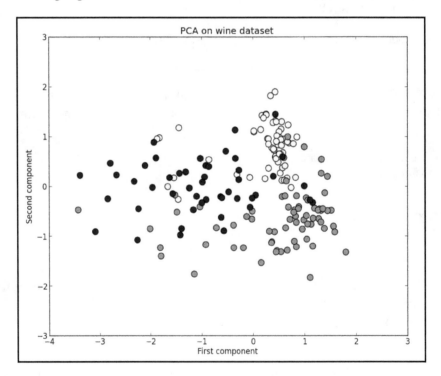

We can try different selections of features and see what the result is. When the distribution of the data is highly dense, we will prefer to select fewer attributes, or mix some of them using proportions or means. In the following diagram, we can see the same implementation using more attributes: `Alcohol`, `Malic acid`, `Ash`, `Alkalinity of ash`, `Magnesium`, and `Total phenols`. Due to this, we can see a different distribution of the points in the scatter plot:

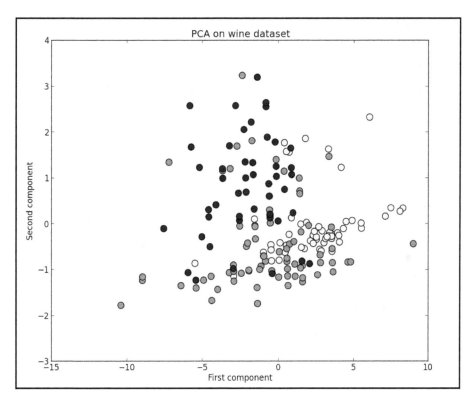

Getting started with SVM

SVM is a supervised classification method based in a kernel geometrical construction, as shown in the following diagram. SVM can be applied for either classification or regression, because a classification problem can be treated as a special type of regression problem, assuming that each observation is placed into one, and only one, of the categories of the values of the predictors. SVM will look for the best decision boundary that splits the points into the classes they belong to. To accomplish this SVM, we will look for the largest margin (space free of training samples parallel to the decision boundary).

In the following diagram, we can see the margin as the space between the dividing line and dotted lines, which extend support vector classifiers to accommodate nonlinear class boundaries. SVM will always look for a global solution because the algorithm only cares about the vectors close to the decision boundary. The points in the edge of the margin are the support vectors. However, this is only for two-dimensional spaces. When we have high-dimensional spaces, the decision boundaries turn into hyper-plane (maximum decision margin). And the SVMs will look for the maximum margin hyper planes. Imagine this from the given labeled training data; the algorithm gives as the result an optimal hyper plane, which categorizes new samples. In this chapter, we will only work with two-dimensional spaces.

 We can find more information and reference of support vector machines and other Kernel-based techniques at `http://www.support-vector-mach ines.org/`.

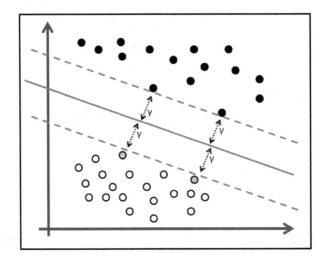

Kernel functions

The linear SVM has two main restrictions. First, the resulted classifier will be linear, and second, we need a dataset that can be split linearly. However, in the real world, many data problems are not linear models. Kernels are a way of creating synthetic variables, which are a combination of individual variables; its goal is to fit data into a space where it can be linearly separable. Due to this, we may want to try with different kinds of kernels. SVMs support many different kinds of kernels, but the most common are as follows:

- **Polynomial**: Polynomial in `mlpy` defined as *kernel_type* = *"poly"*, as shown in the following formula:

$$(<gamma*uT*v>+coef\,0)^{deg\,ree}$$

- **Gaussian**: Gaussian in `mlpy` defined as *kernel_type* = *"rbf"*, as shown in the following formula:

$$\exp\left(\frac{-(u-v)^2}{gamma}\right)$$

- **Sigmoidal**: Sigmoidal in `mlpy` defined as *kernel_type* = *"sigmoid"*, as shown in the following formula:

$$\exp\left(\frac{-(u-v)^2}{gamma}\right)$$

- **Inverse multi-quadratic**: Inverse multi-quadratic is not supported in `mlpy`:

$$\tanh(gamma*uT*v+coef\,0)$$

The double spiral problem

The double spiral problem is a complex artificial problem that tries to distinguish between two classes in a spiral shape. This problem is particularly hard for the classic classifiers, due to its hard mix of values. The datasets are two classes in a spiral, with three turns and 194 points. In the following graph, we use SVMs with a **Gaussian kernel**, and we test the algorithm with different values for gamma. The gamma attribute defines the distance of a single training sample. If the gamma value is low, the attribute is far, and if the value is high, the attribute is close. The algorithm gets better solutions as we increment gamma up to the value of 100:

> For this figure, we used the code and dataset from the `mlpy` reference documentation, which you can find at `http://bit.ly/18SjaiC`.

$$\tanh(gamma * uT * v + coef0)$$

SVM implemented on mlpy

In the following code, we will provide a simple implementation of SVM algorithm with `mlpy`, which implements the **LIBSVM** library. In this case, we will use a linear kernel, assuming that z is the two-dimensional space result of the PCA dimensionality reduction.

First, we need to create a new instance of SVM and define the kernel type as `linear`:

```
svm = mlpy.LibSvm(kernel_type='linear')
```

Then, we will train the algorithm with the `learn` function, using as parameters the two-dimensional space in the variable z and the class that is stored in the variable y:

```
svm.learn(z, y)
```

Now, we need to create a grid where `svm` will perform the predictions in order to visualize the result. We will use the `numpy` functions (`meshgrid`, `arange`) to create the matrix, and then with the revel function, which turns matrices into a list of values for the predictor:

```
xmin, xmax = z[:,0].min()-0.1, z[:,0].max()+0.1
ymin, ymax = z[:,1].min()-0.1, z[:,1].max()+0.1
xx, yy = np.meshgrid(np.arange(xmin, xmax, 0.01),
          np.arange(ymin, ymax, 0.01))
grid = np.c_[xx.ravel(), yy.ravel()]
```

Next, with the `pred` function, we will return the prediction for each point in the grid:

```
result = svm.pred(grid)
```

Finally, we will visualize the predictions in a scatter plot:

```
fig2 = plt.figure(2)
title = plt.title("SVM (linear kernel) on PCA")
plot1 = plt.pcolormesh(xx, yy, result.reshape(xx.shape), cmap=cm.Greys_r)
plot2 = plt.scatter(z[:, 0], z[:, 1], c=y, s=90, cmap=cm.Reds)
labx = plt.xlabel("First component")
laby = plt.ylabel("Second component")
limx = plt.xlim(xmin, xmax)
limy = plt.ylim(ymin, ymax)
plt.show()
```

In the following graph, we can see the result of the plot for the SVM using a linear kernel.

We can observe a clear separation of the three classes. We may also see that the solution does not depend on all points; instead, the separation will depend on only those points that are close to the decision boundary:

$$\frac{1}{\sqrt{(u - gamma)^2 + coef0^2}}$$

In the following graph, we execute the SVM using some more training attributes: `Alcohol`, `Malic acid`, `Ash`, `Alkalinity of ash`, `Magnesium`, and `Total phenols`. Due to this, the plot has different decision boundaries. However, if the SVM can't find a linear separation, the code will get into an infinite loop:

$$\frac{1}{\sqrt{(u - gamma)^2 + coef0^2}}$$

In the following graph, we can see the result of the SVM implementing a Gaussian kernel, and we can observe nonlinear boundaries. The instruction we need to update to get this result is listed here:

```
svm = mlpy.LibSvm(kernel_type='rbf' gamma = 20)
```

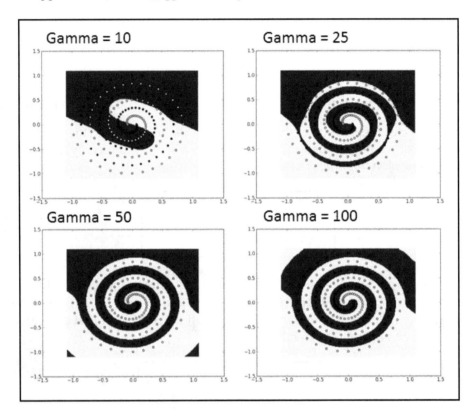

The complete code for Wine classifier using PCA and SVM is listed here:

```
import numpy as np
import mlpy
import matplotlib.pyplot as plt
import matplotlib.cm as cm

wine = np.loadtxt('wine.data', delimiter=',')
x, y = wine[:, 2:5], wine[:, 0].astype(np.int)

pca = mlpy.PCA()
pca.learn(x)
z = pca.transform(x, k=2)
fig1 = plt.figure(1)
```

```
title = plt.title("PCA on wine dataset")
plot = plt.scatter(z[:, 0], z[:, 1], c=y, s=90, cmap=cm.Reds)
labx = plt.xlabel("First component")
laby = plt.ylabel("Second component")
plt.show()

svm = mlpy.LibSvm(kernel_type='linear')
svm.learn(z, y)

xmin, xmax = z[:,0].min()-0.1, z[:,0].max()+0.1
ymin, ymax = z[:,1].min()-0.1, z[:,1].max()+0.1
xx, yy = np.meshgrid(np.arange(xmin, xmax, 0.01),
          np.arange(ymin, ymax, 0.01))
grid = np.c_[xx.ravel(), yy.ravel()]

result = svm.pred(grid)

fig2 = plt.figure(2)
plot1 = plt.pcolormesh(xx, yy, result.reshape(xx.shape), cmap=cm.Greys_r)
plot2 = plt.scatter(z[:, 0], z[:, 1], c=y, s=90, cmap=cm.Reds)
labx = plt.xlabel("First component")
laby = plt.ylabel("Second component")
limx = plt.xlim(xmin, xmax)
limy = plt.ylim(ymin, ymax)
plt.show()
```

All the code and datasets of this chapter can be found in the author's GitHub repository at `https://github.com/hmcuesta/PDA_Book/tree/master/Chapter8`.

Summary

In this chapter, we got into dimensionality reduction and linear classification using SVM. In our example, we created a simple but powerful SVM classifier using different kinds of kernels, and you learned how to perform a dimensionality reduction using PCA implemented in Python with `mlpy`. Finally, we presented how to use nonlinear kernels, such as Gaussian or Polynomial. The work in this chapter was just an introduction to the SVM algorithm, with only two dimensions. The results can be improved with a multidimensional approach with an optimal hyperplane.

In the next chapter, you will learn how to model an epidemiological event (infectious disease) and how an infectious disease is spread through a population. We will create a simulator of an outbreak with a cellular automaton implemented in D3.js.

9
Modeling Infectious Diseases with Cellular Automata

One of the goals of data analysis is to understand the system we are studying, and modeling is the natural way to understand a real-world phenomenon. A model is always a simplified version of the real thing. However, through modeling and simulation, we can try scenarios that are hard to reproduce, or are expensive or dangerous. We can then perform analysis, define thresholds, and provide the information needed to make decisions. In this chapter, we will model an infectious disease outbreak through a **Cellular Automaton (CA)** simulation implemented in JavaScript using D3.js. Finally, we will contrast the results of the simulation with the classical **Ordinary Differential Equations (ODE)**.

This chapter will cover the following topics:

- Introduction to epidemiology
- The epidemic models
- Modeling with cellular automatas
- Implementing a SIRS model in CA with D3js

Introduction to epidemiology

We can define **epidemiology** as the study of the determinants and a distribution of health-related states. We will study how a pathogen, such as the common flu or the influenza AH1N1, is spread within a population. This is particularly important because an outbreak can cause severe human and economic losses, as with the Spanish flu of 1918, which killed 40 million people globally. Take a look at the following screenshot:

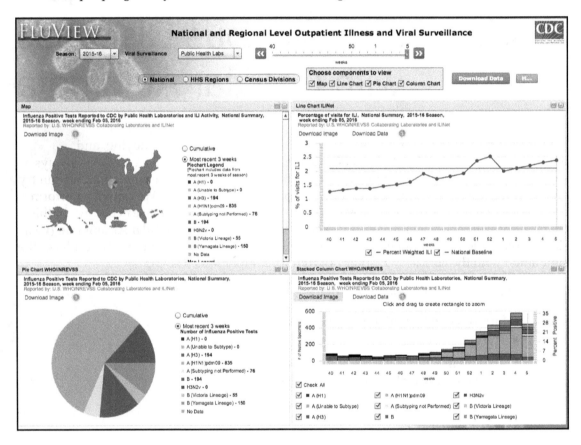

We can use the **Center for Disease Control (CDC)** data, which is freely available from their website. With these time series, we can perform statistical methods for descriptive epidemiology or causal inference. The CDC data is obtained using typical surveys and medical reports, providing real results.

We can use the dashboard for CDC Flu Trends and its data, which is freely available from the following link:

```
http://gis.cdc.gov/grasp/fluview/fluportaldashboard.html
```

Seasonal influenza (flu) data can be found at `http://www.cdc.gov/flu/`.

The epidemiology triangle

In the following diagram, we can see the epidemiologic triangle, which presents all the elements involved in an epidemic outbreak. We can see the **Agent**, which is the infectious pathogen, and the **Host**, which is the human susceptible to acquiring the disease; this will be highly related to its behavior in the environment. The **Environment** includes the external conditions that allow the spread of the disease, such as geography, demography, weather, or social habits. All these elements merge in a **Time** span, and we may see an emerging disease or a seasonal disease:

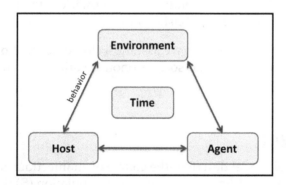

One of the most important concepts to take into account in epidemiology is the **basic reproduction ratio (R-0)**. It is a metric of the number of cases that one infected host can generate in its infectious period. If R-0 is less than 1, it means the infection will vanish in the long run. However, if R-0 is greater than 1, the infection will be able to spread in the host population. We can keep an endemic balance if R-0 in the susceptible population is equal to 1.

The difference between Endemic, Epidemic, and Pandemic is as follows:
Endemic is a disease that exists permanently in a particular population or geographic region.
Epidemic is a disease outbreak that infects many individuals in a population at the same time.
Pandemic is when an epidemic spreads to a worldwide level.
For a complete reference to epidemiology concepts, refer to the textbook *Introduction to Epidemiology 6th Edition*, by Ray M. Merrill, Jones & Bartlett Learning (2012).

The epidemic models

When we want to describe how a pathogen or a disease is spread within a population, we need to create a model using mathematical, statistical, or computational tools. The most common model used in epidemiology is the **SIR** (**Susceptible**, **Infected**, and **Recovered**) model, which was formulated by McKendrick and Kermack in their paper *A Contribution to the Mathematical Theory of Epidemics*, published in 1927.

In the models presented in this chapter, we assume a closed population (without births or deaths) and that the demographics and socio-economic variables do not affect the spread of the disease.

The SIR model

The **SIR** epidemiological model describes the course of an infectious disease, as we can see in the following diagram. Starting with a susceptible population (**S**) that comes into contact with an infected population (**I**), the individual remains infected, and once the infection period has passed, the individual is then in the recovered state (**R**).

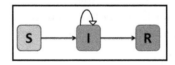

In this chapter, we will use two different ways of solving the SIR model, a mathematical model with an **Ordinary Differential Equations** (**ODE**) system and a computational model using a CA. The two models should show a similar morphology (similar shape of the time series) in the three populations (Susceptible, Recovery, and Recovered) within the timeframe of the outbreak.

Listed here, we can see the **Ordinary Differential Equation** (**ODE**) system that represents the SIR model:

$$\text{(a)} \quad \frac{dS}{dt} = -\beta * S * I$$

$$\text{(b)} \quad \frac{dI}{dt} = \beta * S * I - \gamma * I$$

$$\text{(c)} \quad \frac{dR}{dt} = \gamma * I$$

Solving the ordinary differential equation for the SIR model with SciPy

In order to observe the morphology of an infectious disease outbreak, we need to solve the SIR model. In this case, we will use the `integrate` method of the SciPy module to solve the ODE. First, we need to import the required libraries, `scipy` and `pylab`:

```
import scipy
import scipy.integrate
import pylab as plt
```

Then, we will define the `SIR_model` function, which will contain the ODE, with beta representing the transmission probability, gamma as the infected period, `X[0]` representing the susceptible population, and `X[1]` the infected population:

```
beta = 0.003
gamma = 0.1

def SIR_model(X, t=0):
    r = scipy.array([- beta*X[0]*X[1]
                    , beta*X[0]*X[1] - gamma*X[1]
                    , gamma*X[1] ])
    return r
```

Next, we will define the initial `parameters` (SIR) and the `time` (number of days), then, with the `scipy.integrate.odeint` function, we will solve the differential equations system:

```
if __name__ == "__main__":

        time = scipy.linspace(0, 60, num = 100)
        parameters = scipy.array([225, 1,0])
        X = scipy.integrate.odeint(SIR_model, parameters,time)
```

The result of the `SIR_model` will look like the following list and will contain the status of three populations (SIR) for each step (days) during the outbreak:

```
[[  2.25000000e+02   1.00000000e+00   0.00000000e+00]
 [  2.24511177e+02   1.41632577e+00   7.24969630e-02]
 [  2.23821028e+02   2.00385053e+00   1.75121774e-01]
 [  2.22848937e+02   2.83085357e+00   3.20209039e-01]
 [  2.21484283e+02   3.99075767e+00   5.24959040e-01]
 . . .]
```

All the codes and datasets of this chapter may be found in the author's GitHub repository at:
`https://github.com/hmcuesta/PDA_Book/tree/master/Chapter9`

Finally, we will plot the three populations using `pylab`:

```
plt.plot(range(0, 100), X[:,0], 'o', color ="green")
plt.plot(range(0, 100), X[:,1], 'x', color ="red")
plt.plot(range(0, 100), X[:,2], '*', color ="blue")
plt.show()
```

In the following screenshot, we can see the transition rates for the SIR model:

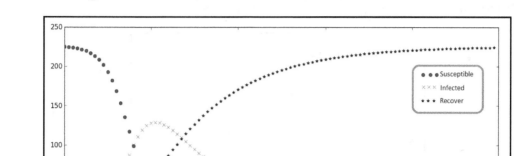

The SIRS model

The **SIRS** (**Susceptible, Infected, Recover,** and **Susceptible**) model is an extension of the SIR model. In this case, the immunity acquired in the recovered status is eventually lost and the individual eventually comes back to the susceptible population. As we can see in the following diagram, the SIRS model is cyclical. The SIRS model brings the opportunity to study other kinds of phenomenon, such as endemic diseases and seasonality effects. Some common examples of SIRS diseases are **Seasonal Flu, Measles, Diphtheria,** and **Chickenpox**:

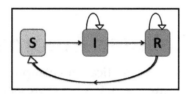

Listed here, we can see the **Ordinary Differential Equation** (ODE) system that represents the SIRS model:

$$
\begin{aligned}
&\text{(a)} \quad \frac{dS}{dt} = -\beta * S * I + \sigma * R \\[2mm]
&\text{(b)} \quad \frac{dI}{dt} = \beta * S * I - \gamma * I \\[2mm]
&\text{(c)} \quad \frac{dR}{dt} = \gamma * I - \sigma * R
\end{aligned}
$$

In order to solve the ODE (see the section *Solving ordinary differential equations for the SIR model with SciPy*), we need to create the `SIRS_model` function as listed next, where the sigma variable represents the recovered period, as is shown in the ODE of the SIRS, using the `beta` variable to represent the transmission probability, and `gamma` as the infected period. Finally, we will use `X[0]` to represent the susceptible population, with `X[1]` as the infected population and `X[2]` as the recovered population:

```
beta = 0.003
gamma = 0.1
sigma = 0.1

def SIRS_model(X, t=0):
    r = scipy.array([- beta*X[0]*X[1] + sigma*X[2]
                    ,  beta*X[0]*X[1] - gamma*X[1]
                    ,  gamma*X[1] ] -sigma*X[2])
    return r
```

Modeling with Cellular Automaton

Cellular Automaton (CA) are the mathematical and computational discrete models created by John von Neumann and Stanislaw Ulam. CA are represented as a grid where, in each cell, a small computation is performed. In CA, we will share the process throughout all the small cells in the grid and the CA shows a behavior similar to biological reproduction and evolution. In this case, we may say that each cell is an individual in our population (grid) that will switch between states depending on its social interaction (contact rate; see SIR and SIRS models).

Seen as discrete simulations of dynamic systems, the CA has been used for modeling in different areas, such as traffic flow, encryption, growth of crystals, bird migration, and epidemic outbreaks. Stephen Wolfram, one of the most influential researchers in CA, describes CA as:

> *"Cellular automata are sufficiently simple to allow detailed mathematical analysis, yet sufficiently complex to exhibit a wide variety of complicated phenomena."*

Cell, state, grid, neighborhood

The basic element in a CA is the **cell** , which corresponds to a specific coordinate in a **grid** (or lattice). Each cell has a finite number of possible states and the current**state** will depend on a set of rules and the status of the surrounding cells (**neighborhood**). All the cells follow the same set of rules, and when the rules are applied to the entire grid, we can say that a new generation is created.

There are different kinds of neighborhood, as we can see in the following diagram:

- **Von Neumann**: This encompasses the four cells orthogonally surrounding a central cell on a two-dimensional square grid.
- **Moore**: This is the most common neighborhood and encompasses the eight cells that surround the central cell in a two-dimensional grid.
- **Moore Extended**: This has basically the same behavior as the Moore, but in this case, we can extend the reach to different distances.
- **Global**: The geometric distance is not considered, and in this case, all the cells have the same probability of being reached by another cell. (See the *Global stochastic contact model* section.)

One of the most famous examples of CA is Conway's **Game of Life**, where, in a two-dimensional lattice, all the cells can be either dead or alive. In this link, we can see a D3.js visualization of the Game of Life: http://bl.ocks.org/sylvaingi/2369589

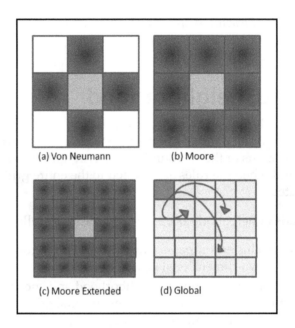

(a) Von Neumann   (b) Moore

(c) Moore Extended   (d) Global

Global stochastic contact model

For this model, we will define the interaction between individuals in a homogeneous population. The contact is global and stochastic; this means that each cell has the same likelihood of being contacted by another. In this model, we do not consider the geographic distance, demographics, or the migratory pattern as constraints.

You can find more information about the stochastic process at: http://en.wikipedia.org/wiki/Stochastic_process

Simulation of the SIRS model in CA with D3.js

In Chapter 7, *Predicting Gold Prices*, we already studied the basics of a random walk simulation. In this chapter, we will implement a CA in JavaScript using D3.js to simulate the SIRS model.

In the following screenshot, we can see the interface of our simulator. It's a simple interface with a 15×15 grid of cells (225 cells in total). One button, **Update**, applies the rules to all the cells on the grid (step). One paragraph area will show the status of different populations in the current step, for example, Susceptible: 35 Infected: 153 Recovered: 37 Step: 4. Finally, a button named **Statistics** writes a list with all the stats of each step [Susceptible, Infected, Recovered ….] into a text area for the purpose of plotting.

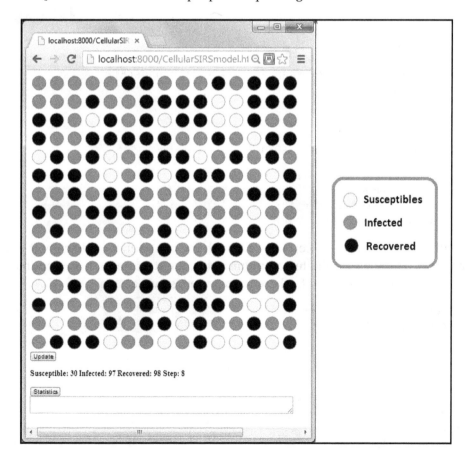

Inside the `head` tag, we need to refer to the library:

```
<html>
<head>
  <script src="http://d3js.org/d3.v3.min.js"></script>
</head>
```

The code is mostly JavaScript; first, we need to define the variables such as the grid, the list of `colors`, and the number of `rows` and `columns`, and then the SIRS model parameters, such as the average number of contacts (`avgContact`), transmission probability (`tProb`), initial number of infected (`initialInfected`), and infected and recovered periods (`timeRecover` and `timeInfection`):

```
<body>
  <script type="text/javascript">

    var w = 600;
    var h = 600;
    var grid = [];
    var record = [];
    var colors = ["", "#F8F8F8", "#FF6633","000066"];
    var index = 0;
    var cols = 15;
    var rows =15;
    var nTimes = 0;
    var gridSize = (cols * rows);
    var avgContact = 4;
    var timeInfection = 2;
    var timeRecover = 4;
    var initialInfected = 10;
    var tProb = 0.2;
```

Now we will fill the grid with susceptible cells with the `push` function. Each of the cells will contain an array with the coordinates, a unique index, the start status 1 (susceptible), and the period of time (in the initial state, there is no period, so we assign):

```
for(var i=0; i <rows; i++ ){
  for(var j=0;j<cols;j++){
    grid.push([i*40,
                       j*40,
                       "circle-"+index++,
                       1,
                       0]);
      }
    }
```

Then, we need to define the size of the new SVG's width and height (600×600 pixels); this inserts a new <svg> element before the closing </body> tag:

```
var svg = d3.select("body")
    .append("svg")
    .attr("width", w)
    .attr("height", h)
    .append("g")
    .attr("transform","translate(20,20)");
```

Next, we need to generate the circle elements and add them to svg, then with the data(grid) function, for each value in the data, we will call the .enter() function and add a "circle" element. D3 allows selecting groups of elements for manipulation through the selectAll function. We will use each cell array inside the grid list to define the coordinates (cx, cy), color, and id:

```
svg.selectAll("circle")
    .data(grid)
    .enter()
    .append("circle")
    .attr("id", function(d) {
      return d[2];
    })
    .attr("cx", function(d) {
        return d[0];
    })
    .attr("cy", function(d) {
        return d[1];
    })
    .attr("r", function(d) {
        return 15;
    })
    .attr("fill", colors[1])
    .attr("stroke", "#666");
```

Next, we will create the init function; this will be called only when we refresh the web page. init will randomly insert the initial number of infected cells in the CA:

```
function init(){
    for(var x = 0; x < initialInfected; x++ ){
    var i = Math.round(Math.random() * (gridSize-1));
    var cell = grid[i];
    if(cell[3]==1){
      cell[3] = 2;
      cell[4] = timeInfection;
    }
    grid[i] = cell;
```

```
      }
    prepareStep();
    }
  init();
```

In the `prepareStep` function, we will refill all the circles with their new status colors (`colors[cell[3]]`). We will use the `svg.select` function to select one element by its id (`cell[2]`) and apply the new style. `prepareStep` also counts the number of individuals in each of the three populations and shows them in the paragraph tag (`status`). Finally, the function stores the statistics of the current step in the `record` list:

```
function prepareStep(){
noSus = 0;
  noInfected = 0;
  noRecover = 0;
  nTimes++;
  for(var i = 0; i < gridSize; i++){
    var cell = grid[i];
    svg.select("#"+cell[2]).style("fill", colors[cell[3]]);
    if(cell[3] == 1){
      noSus++;
    }else if(cell[3] == 2){
      noInfected++;
    }else if(cell[3] == 3){
      noRecover++;
    }
  }
  record.push([noSus,noInfected,noRecover]);
  document.getElementById("status").innerHTML =
                          " Suseptibles: "+ noSus+
                          " Infected: "+noInfected+
                          " Recovered: "+noRecover+
                          " Times: "+ nTimes;

}
```

In `nextStep`, we will apply the rules defined in the SIRS model to each cell to define their new statuses. We will use the relative ID of the cells instead of their coordinates, as it will be easier to reach the cell by its position in the list (0 to 224):

```
function nextStep(){
    for(var i=0; i < gridSize; i++){
```

We will take each `cell` one by one, and we will apply their average number of contacts with the other cells:

```
    var cell = grid[i];
```

We will check whether the cell is in the *Recovered* status (3); if the cell still has a time period in this status, we just decrement the recovered period by 1. However, if the recovered period is , we will perform the transition to the Susceptible status (1):

```
if(cell[3]==3){
  if(cell[4] > 0){
    cell[4] = cell[4] - 1;
            }else{
    cell[3] = 1;
    cell[4] = 0;
  }
}else{
```

Now, if the status is 1 or 2 (Susceptible or Infected), we need to make random contacts and compare the status of the first `cell` with the status of the second cell (`sCell`). If they have the same status, continue with the next contact. If either of them is infected, the other cell is exposed to the transmission probability (`tProb`), and if it's infected, the cell is updated in the grid:

```
for(var j=0;j < avgContact ;j++){
    var sId = Math.round(Math.random() *
                            (gridSize-1));
    var sCell = grid[sId];
    if(cell[3] == sCell[3]){
      continue;
    }else if (cell[3] == 2 && sCell[3] == 1){
      if(Math.random() <= tProb){
        sCell[3] = 2;
        sCell[4] = timeInfection;
      }
    }else if (cell[3] == 1 && sCell[3] == 2){
      if(Math.random() <= tProb){
        cell[3] = 2;
        cell[4] = timeInfection;
      }
    }
    grid[sId] = sCell;
  }
}
```

Next, if the cell is in the status infected (2), we will check whether the period is over. In this case, we perform the transition to the recovered status. Otherwise, we just decrease the timer of the infectious period (`cell[4]`):

```
if(cell[3] == 2 && cell[4] == 0 ){
  cell[3] = 3;
  cell[4] = timeRecover;
```

```
    }else if(cell[3] == 2 && cell[4] > 0 ){
        cell[4] = cell[4] - 1;
    }
    grid[i] = cell;
    }
}
```

The `update` function triggers the new step for the CA by calling the `nextStep` and `prepareStep` functions:

```
function update(){
    nextStep();
    prepareStep();
}
```

The `statistics` function writes the `record` list with the statistics of the simulation into the text area tag (`txArea`):

```
function statistics(){
    document.getElementById("txArea").value = ""+record;
}
</script>
```

Finally, we create the entire HTML needed for the interface, the `"Update"` button, paragraph area (`status`), the `"Statistics"` button, and the text area (`txArea`):

```
<div id="option">
<input name="updateButton"
        type="button"
        value="Update"
        onclick="update()" />
</div>
<p id="status">Current Statistics</p>
<input name="updateButton"
        type="button"
        value="Statistics"
        onclick="statistics()" />
</br>
<textarea id=txArea
        cols = "70">
</textarea>
</body>
</html>
```

In the following screenshot, we can observe the progression of the outbreak in **Step 1**, **Step 3**, **Step 6**, **Step 9**, **Step 11**, and **Step 14**. We can appreciate how the SIRS model is applied to the grid:

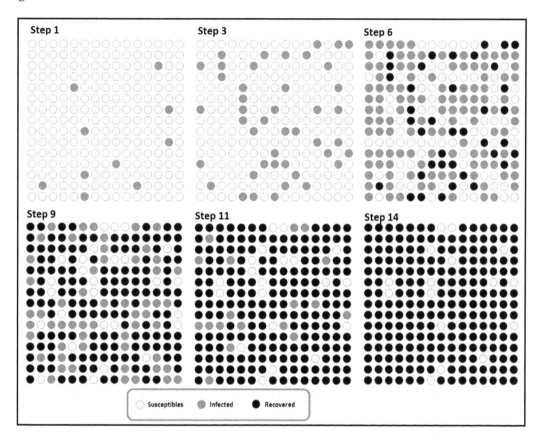

Now, we will copy the `record` list from the text area and we will visualize it in Python with the small script listed later.

First, we import the `pylab` and `numpy` modules:

```
import pylab as plt
import numpy as np
```

Then, we will create a `numpy` array with the record list:

```
data = np.array([215,10,. . .])
```

Next, in order to plot each population, we will reshape the `array` with `numpy`. We will now use the `reshape` method; the first parameter is (-1) because we don't know in advance how many steps there are, and the second parameter is to define the length of the array in three (SIR):

```
result = data.reshape(-1,3)
```

The resulting array will look like this:

```
[[215  10   0]
 [153  72   0]
 [ 54 171   0]
 [  2 223   0]
 [  0 225   0]
 [  0 178  47]
 [  0  72 153]
 [  0   6 219]
 [  0   0 225]
 [ 47   0 178]
 [153   0  72]
 [219   0   6]
 [225   0   0]]
```

Finally, we use the `plot` method to display the visualization:

```
length = len(result)
plt.plot(range(0,length), result[:,0], marker = 'o', lw = 3, color="green")
plt.plot(range(0,length), result[:,1], marker = 'x', linestyle = '--', lw =
3, color="red")
plt.plot(range(0,length), result[:,2], marker = '*', linestyle = ':',lw =
3, color="blue")
plt.show()
```

In the following diagram, we can see the three populations through time, until all the cells return to **Susceptible**:

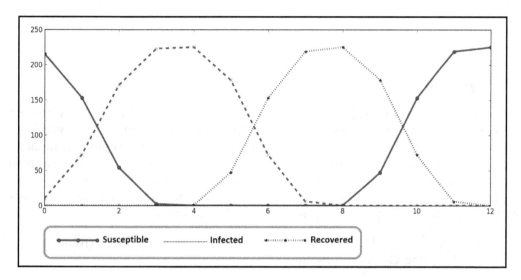

We may also play with the parameters, such as the infectious period, the initial number of infected individuals, the transmission probability, or the recovery period. In the following diagram, we simulate the SIR model by increasing the recovery period to a large number and as we can observe, the result is highly similar to the result given by the mathematical model (ODE). See the *Solving ordinary differential equation for the SIR model with SciPy* section.

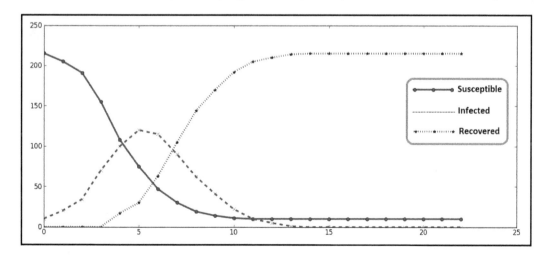

All the code and datasets for this chapter may be found in the author's GitHub repository at:
`https://github.com/hmcuesta/PDA_Book/tree/master/Chapter9`

Summary

In this chapter, we introduced the basic concepts of Epidemiology and two basic epidemic models (SIR and SIRS) for infections transmitted from human to human. Then, you learned how to model and solve an ordinary differential equation's system for epidemic models. Finally, we developed a basic simulator implementing a CA of the SIRS model in a small population. We try different parameters and get interesting results. Of course, these examples are only for educational purposes and if we need to model a real disease, we will need an epidemiologist to provide accurate and real parameters.

In the next chapter, you will learn how to visualize and work with graphs from social networking sites using **Gephi** and **Python**.

10
Working with Social Graphs

In this chapter, we will introduce the most basic features of **graph analytics**. Initially, we will distinguish the structure of a graph and a social network analysis. Then, we will present some of the basic operations with a graph, such as **Degree** and **Centrality**. Then, we will work in a graph representation using a visualization tool called **Gephi** to apply some statistical and visualization methods implemented on graphs, such as modularity classification, coloring nodes, and layout distributions. Finally, we will create our own visualization in D3.js for our friends graph.

This chapter will cover the following:

- Social networks analysis
- Working with graphs using Gephi
- Statistical analysis of my graph (Degree and Centrality)
- Graph visualization with D3.js

Structure of a graph

A graph is a set of **Nodes** (or Vertices) and a set of **Links** (or Edges). Each link is a pair of node references (**Source** and **Target**). Links may be considered **directed** or **undirected**, if the relationship is one way or two ways. The most common way to computationally represent a graph is using an adjacency matrix with the index of the matrix as a node identifier and a value in the coordinates to represent whether a link exists (1) or not (0). The links between nodes may have a scalar value (weight) to define a distance between nodes. Graphs are widely used in Sociology, Epidemiology, Internet, Government, Commerce, and Social Networks to find groups and for information diffusion.

Undirected graph

In an undirected graph, there is no distinction between the node's source and target. As we can observe in the following diagram, the adjacency matrix is symmetrical, which means that the relationship between nodes is mutual. This is the kind of graph used in Facebook, where we are friends with other nodes (symmetrical relationship):

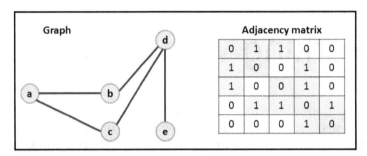

Directed graph

In a directed graph, we find direction between the source node and the target node represented by an arrow; this creates an asymmetrical (one-way) relationship. In this case, we will have two different kinds of degree: *In* and *Out*. This can be observed in the adjacency matrix, which is not symmetrical. This is particularly useful in networks such as Twitter or Google+, where we have followers, not friends. This means that the relationship is not mutual by default and we will have two degrees: **In (Followers)** and **Out (Following)**:

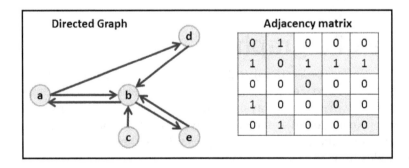

Social networks analysis

Social Networks Analysis (**SNA**) is not new; sociologists have been using it for a long time to study human relationships (**sociometry**), find communities, and simulate how information or a disease is spread in a population.

With the rise of social networking sites such as Facebook, Twitter, LinkedIn, and so on, the acquisition of large amounts of social network data is easier. We can use SNA to get insight about customer behavior or unknown communities. It is important to say that this is not a trivial task, and we will face sparse data and a lot of noise (meaningless data). We need to understand how to distinguish between false correlation and causation. A good start is to know our graph through visualization and statistical analysis.

Social network sites give us opportunities to ask questions that are otherwise too hard to approach, because pooling enough people is time-consuming and expensive.

In this chapter, you will learn how to get insight into the proportions of the non-graph data provided by Facebook, such as by gender or likes. Next, we will explore the distribution and centrality of our friend relationships in our graph. Finally, we will create an interactive visualization of our graph using D3.js.

Acquiring the Facebook graph

In Facebook, friends are represented by nodes and the relationship between friends is represented by links, but we can get a lot more information from it, such as gender, age, post list, likes, political affiliation, religion, and so on, and Facebook provides us with a complete HTTP-based **API** (**Application Programming Interface**) to work with its data. Follow this link for more information:

```
https://developers.facebook.com/
```

Another interesting option is the **Stanford Large Network Dataset Collection**, where we can find social network datasets well-formatted and anonymized for educational proposes. Follow this link for more information:

```
http://snap.stanford.edu/data/
```

Using anonymized data, it is possible to determine whether two users have the same affiliations, but not what their individual affiliations represent.

In this chapter, we will use a Facebook graph with 1,274 friends and 43,000 relationships between them. This will help us understand how friends from social networks works and we may find patterns and groups in them.

To download the datasets used in this chapter, please follow this link: `https://github.com/hmcuesta/Datasets/`

The dataset file `firnds.gdf` will look like this:

```
nodedef>name VARCHAR,label VARCHAR,gender VARCHAR,locale
VARCHAR,agerank INT
23917067,Jorge,male,en_US,106
23931909,Haruna,female,en_US,105
35702006,Joseph,male,en_US,104
503839109,Damian,male,en_US,103
532735006,Isaac,male,es_LA,102

 .  .  .

edgedef>node1 VARCHAR,node2 VARCHAR
23917067,35702006
23917067,629395837
23917067,747343482
23917067,755605075
23917067,1186286815

 .   .   .
```

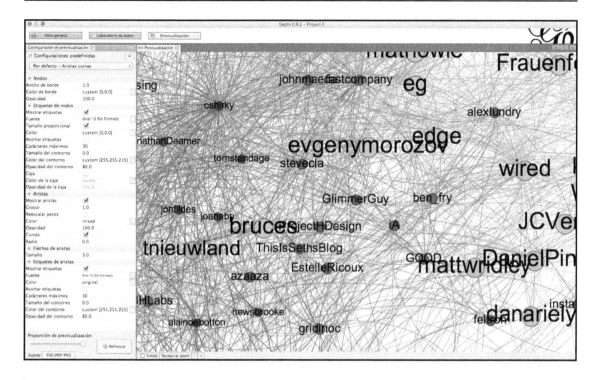

Working with graphs using Gephi

Gephi is open source software for visualizing and analyzing large networks graphs, which runs on Windows, Linux, and Mac OS X. We can freely download Gephi from its website, listed here. For installation instructions, please refer to `Appendix`, *Setting Up the Infrastructure*:

```
https://gephi.org/users/download/
```

To visualize your social network graph, you just need to open Gephi, click on the **File** menu and select **Open**, then look up and select our file, `friends.gdf`, and click on the **Open** button. We can see our graph in the following screenshot.

In the following screenshot, we can see the visualization of the graph (**1,274** nodes and **43,928** links). The nodes represent friends and the links represent how the friends are connected to each other. The graph looks very dense and does not provide us with much insight; the next step is to find groups through a **Modularity algorithm** and color classification:

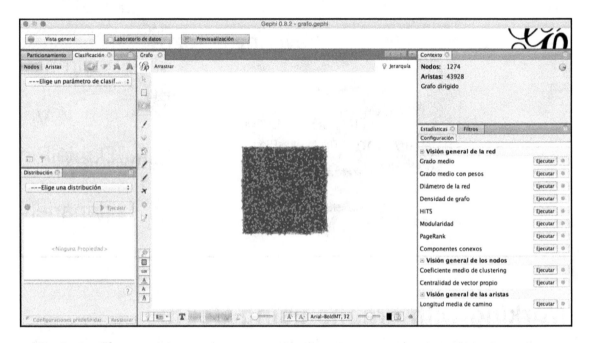

For complete reference documentation on Gephi, please refer to this link: `https://gephi.org/users/`

In the interface, we can see the **Context** tab, which shows us the number of **Nodes** and **Edges**. We can show Node labels by clicking on the **T** icon at the bottom of the window. To understand the real nature of the node organization in a graph, we need to detect communities such as family, classmates, close friends, and so on. Gephi implements the Modularity algorithm created by Vincent D. Blondel to quickly unfold communities in large networks; this is going to group the friends into clusters. Take a look at the following screenshot:

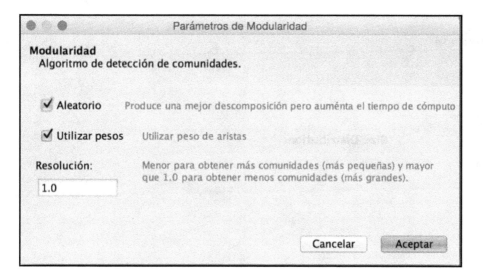

In the **Statistics** panel, we click on **Modularity**; this will generate a **Modularity Report,** as shown in the following screenshot, where we can see the number of communities, the distribution, and the modularity rate:

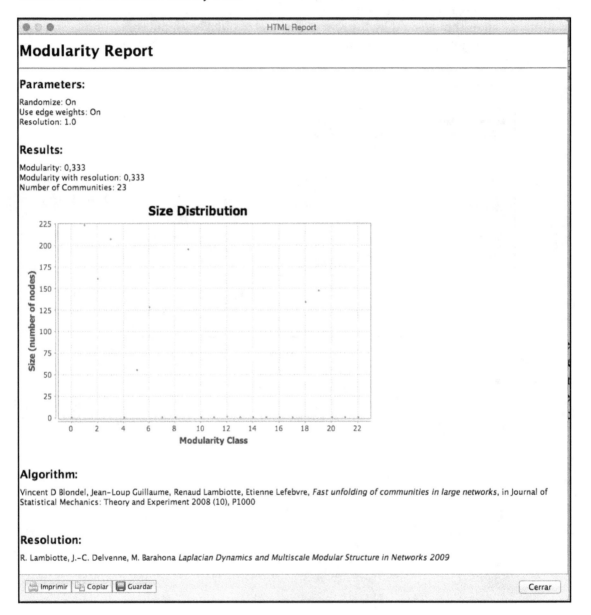

The next step is to generate a color classification based on the Modularity we already made. We need to go to the **Classification** panel and select **Modularity Class**, as we can see in the following screenshot. Then, we can select the color range for the communities and click on **Apply**. This will change the color of the nodes, but it still looks very dense, so we need to apply a layout distribution:

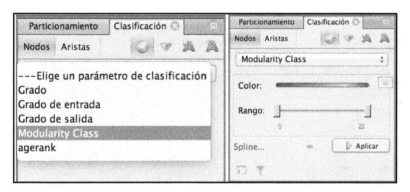

Then, we can apply different layout algorithms by selecting **Choose a layout** in the **Layout** tab. In this case, we can use different layouts to get more understandable images, and we can start with a **Yifan Hu Proporcional**, which is commonly used in recommendation systems for graph drawing, and then use *expansion* to get better details of the community's separation through node overlap removal:

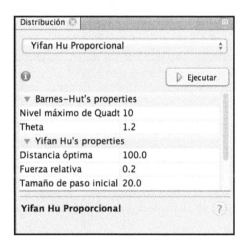

In the following screenshot, we see the preview visualization of the graph using the **Fruchterman-Reingold** algorithm, which is a force-directed layout algorithm. The force-directed layouts are a family of algorithms for drawing graphs in dimensional spaces (2D or 3D) in order to represent the nodes and links of the graph in an aesthetical way:

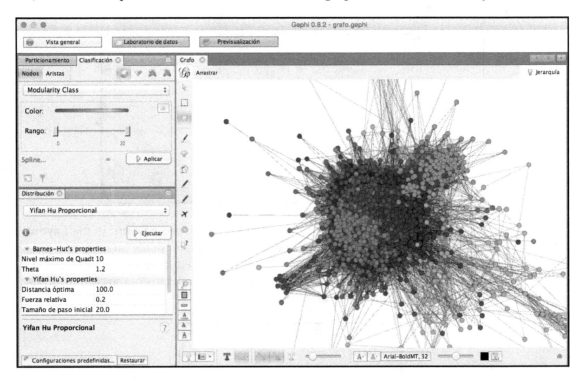

For more information about the Fruchterman-Reingold layout algorithm, follow this link:
`http://wiki.gephi.org/index.php/Fruchterman-Reingold`

Once the visualization is ready, we can click on the **Preview** button to get a better look at it and we can **Export** the visualization to PDF, SVG, or PNG formats:

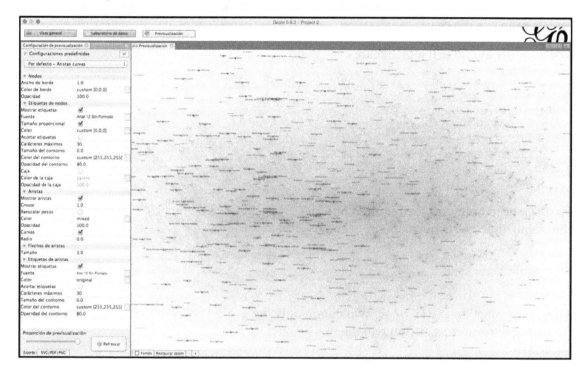

Statistical analysis

We can easily find information, such as the number of friends and individual data of each one, from our Facebook graph. However, there are many answers that we can't get directly from the site, for example, *male to female ratio* or *How many of my friends are Republicans?*, or *who is my best friend?*; these questions can be easily answered with a few lines of code and some basic statistical analysis. In this chapter, we will start with the *male to female ratio*, because we already have the gender value in the `.gdf` file.

For simplicity in the code examples, we will split the `friends.gdf` file into two CSV files—one for the nodes (`nodes.csv`) and one for the links (`links.csv`).

Male to female ratio

In this example, we will use the gender value of the `nodes.csv` file and get the male to female ratio in a pie chart visualization.

The `nodes.csv` file will look like this:

```
nodedef>name VARCHAR,label VARCHAR,gender VARCHAR,locale VARCHAR,agerank
INT
23917067,Jorge,male,en_US,106
23931909,Haruna,female,en_US,105
35702006,Joseph,male,en_US,104
503839109,Damian,male,en_US,103
532735006,Isaac,male,es_LA,102
. . .
```

First, we need to import the required libraries. See `Appendix`, *Setting Up the Infrastructure*, for the installation instructions of `numpy` and `pylab`:

```
import numpy as np
import operator
from pylab import *
```

The `numpy` function `genfromtxt` will obtain only the gender column from the `nodes.csv` file, using the `usecols` attribute in the `str` format:

```
nodes = np.genfromtxt("nodes.csv",
                      dtype=str,
                      delimiter=',',
                      skip_header=1,
                      usecols=(2))
```

Then, we will use the `countOf` function from the `operator` module and ask for how many males are in the list `nodes`:

```
counter = operator.countOf(nodes, 'male')
```

Now, we just get the proportions of `male` and `female` in percentages:

```
male = (counter *100) / len(nodes)
female = 100 - male
```

Next, we make a square `figure` and `axes`:

```
figure(1, figsize=(6,6))
ax = axes([0.1, 0.1, 0.8, 0.8])
```

Then, the slices will be ordered and plotted counterclockwise:

```
labels = 'Male', 'Female'
ratio = [male,female]
explode=(0, 0.05)
```

With the `pie` function, we define the parameters of the chart, such as `explode`, `labels`, and `title`:

```
pie(ratio,
    explode=explode,
    labels=labels,
    title('Male to Female Ratio',
        bbox={'facecolor':'0.8', 'pad':5})
```

Finally, with the `show` function, we execute the visualization:

```
show()
```

In the following diagram, we can see the pie chart and observe that 54.7 percent are male and 45.3 percent are female:

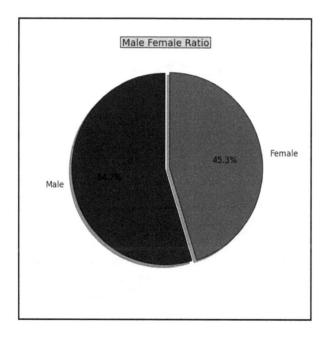

Degree distribution

The degree of a node is the number of connections (links) with other nodes. In the case of directed graphs, each node has two degrees—the out-degree and the in-degree. In the undirected graph, the relationship is mutual; then we just have a single degree for each node. In the code listed here, we get the source node and target node references from the `links.csv` file. Then, we create a single list merging the two lists (target and source). Finally, we get a dictionary (dic) of how many times each node appears in the list, and we plot the result in a bar char using `matplotlib`.

The `links.csv` file will look like this:

```
edgedef>node1 VARCHAR,node2 VARCHAR
23917067,35702006
23917067,629395837
23917067,747343482
23917067,755605075
23917067,1186286815
.  .  .
import numpy as np
import matplotlib.pyplot as plt

links = np.genfromtxt("links.csv",
                      dtype=str,
                      delimiter=',',
                      skip_header=1,
                      usecols=(0,1))
dic = {}
for n in sorted(np.reshape(links,558)):
    if n not in dic:
        dic[n] = 1
    else:
        dic[n] += 1
plt.bar(range(95),list(sort.values()))
plt.xticks(range(95), list(sort.keys()), rotation=90)
plt.show()
```

In the following diagram, we can observe the degree of each node in the graph and there are 11 nodes that do not have any connections. In this example, from the 106 total nodes in the graph, we only consider the 95 nodes that have at least a degree of one:

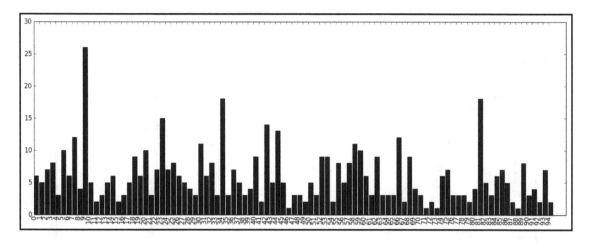

Histogram of a graph

Now, we will explore the structural task of the graph through its histogram. We will create a dictionary (`histogram`) that will contain information about how many nodes have the degree of one, two, or 26, which is the maximum degree that can be reached by a node in this graph. Then, we will visualize the `histogram` using a scatter plot:

```
histogram = {}

for n in range(26):
    histogram[n] = operator.countOf(list(dic.values()), n)

plt.bar(list(histogram.keys())),list(histogram.values()))
plt.show()
```

In the following diagram, we can see the histogram of our graph. The logical question here is *What does the histogram tell us about the graph?* The answer is that we can see a pattern in the histogram and that many nodes in the graph have a small degree, which decreases as we move along the **X**-axis; we can observe that most nodes have a degree of 3. In the pattern, we can appreciate that it becomes less likely that a new node comes with a high degree, which is congruent with the **Zipfian** distribution. Most human-generated data presents this kind of distribution, such as words in a vocabulary, letters in the alphabet, and so on.

For more information about Zipfian distribution, follow this link:
`http://en.wikipedia.org/wiki/Zipf's_law`

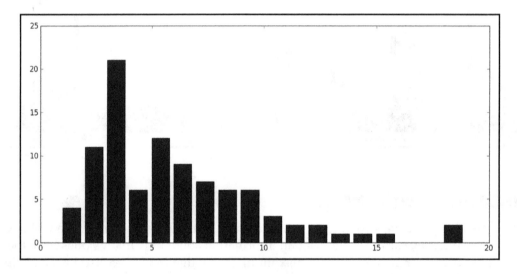

Another common pattern in graphs is the **Exponential Distribution**, which is frequently presented in random graphs.

Centrality

If we want to understand the importance of an individual node in the graph, we need to define its centrality, which is a relative measure of how important a node is within a graph. There are several ways to find a centrality, such as **Closeness** (the average length of all its shortest paths) or **Betweenness** (the fraction of all shortest paths that pass through a certain node). In this case, we will define centrality as the strongest connected node and we will prove this hypothesis through a direct data exploration.

In the code listed here, we sort the dictionary by its value using a lambda function, then we reverse the order to get the biggest degree first:

```
sort = sorted(dic.items(), key=lambda x: x[1], reverse=True)
```

The resulting list, `sort`, will look like this:

```
[('100001448673085', 26),
 ('100001452692990', 18),
 ('100001324112124', 18),
 ('100002339024698', 15),
 ('100000902412307', 14),
 .  .  . ]
```

In the following screenshot, we can see the graph visualized in Gephi with the **Yifan Hu Layout** algorithm. With a direct data exploration, we can color the node with the apparent highest degree and we can say that is the central node. In the Gephi interface, we will select the **Ranking** tab, choose the **Degree** option in the combo box, pick from **Color**, select the highest **Range** (26/27), and click on the **Apply** button (see the underlined options in the following screenshot). We can also color the first degree contacts of the central node and see that it is strongly connected between groups. We can do this by selecting the Painter tool in Gephi and clicking on all the nodes related to the central node. In fact, we find that it is the same node, with the highest degree obtained by the sorted process (ID 100001448673085):

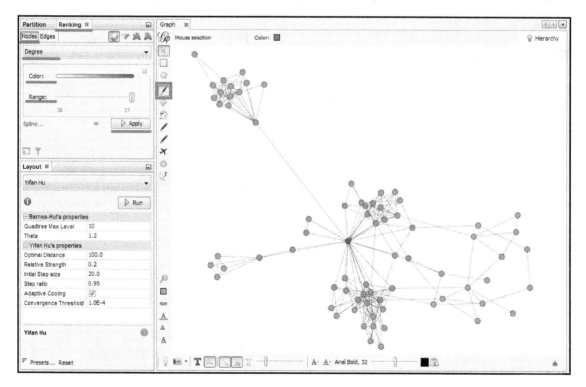

We can create our own centrality algorithm that is not based only on the degree (number of connections of the node). For example, we can find centrality in the number of shares and likes of certain nodes' posts. This means that a node with a lower degree may have a bigger impact on the information diffusion process, or that a particular node is strongly connected between different groups—that's the beauty of social networks.

Transforming GDF to JSON

Gephi is an excellent tool to get easy and fast results. However, if we want to present the graph interactively on a website, we will need to implement a different kind of visualization. In order to work with the graph in the web, we need to transform our `.gdf` file to the JSON format.

First we need to import the libraries `numpy` and `json`. See `Chapter 2`, *Preprocessing the Data*, for more information about the JSON format:

```
import numpy as np
import json
```

The numpy function `genfromtxt` will obtain only the ID and Name from the `nodes.csv` file, using the `usecols` attribute in the `'object'` format:

```
nodes = np.genfromtxt("nodes.csv",
                      dtype='object',
                      delimiter=',',
                      skip_header=1,
                      usecols=(0,1))
```

Then, the numpy function `genfromtxt` will obtain links with the source node and target nodes from the `links.csv` file, using the `usecols` attribute in the `'object'` format:

```
links = np.genfromtxt("links.csv",
                      dtype='object',
                      delimiter=',',
                      skip_header=1,
                      usecols=(0,1))
```

The JSON format used in the D3.js **Force Layout** graph implemented in this chapter requires transforming the ID (ex. 100001448673085) into a numerical position in the list of nodes. Then we need to look for each appearance of the ID in the links and replace them with the position in the list of nodes:

```
for n in range(len(nodes)):
    for ls in range(len(links)):
        if nodes[n][0] == links[ls][0]:
            links[ls][0] = n
        if nodes[n][0] == links[ls][1]:
            links[ls][1] = n
```

Now, we need to create a `data` dictionary to store the JSON:

```
data ={}
```

Next, we need to create a list of nodes with the names of friends in the format `"nodes"`:
`[{"name": "X"},{"name": "Y"},. . .]` and add it to the `data` dictionary:

```
lst = []
for x in nodes:
    d = {}
    d["name"] = str(x[1]).replace("b'","").replace("'","")
    lst.append(d)
data["nodes"] = lst
```

Now, we need to create a list of links with the source and target in the format `"links"`:
`[{"source": 0, "target": 2},{"source": 1, "target": 2},. . .]` and add it to
the `data` dictionary:

```
lnks = []

for ls in links:
    d = {}
    d["source"] = ls[0]
    d["target"] = ls[1]
    lnks.append(d)

data["links"] = lnks
```

Finally, we need to create the `newJson.json` file and write the data dictionary in the file
with the `dumps` function of the `json` library:

```
with open("newJson.json","w") as f:
    f.write(json.dumps(data))
```

Neo4j is a robust (fully ACID) transactional property graph database.
Follow this link for more information about Neo4j:
`http://www.neo4j.org/`

The `newJson.json` file will look like this:

```
{"nodes": [{"name": "Jorge"},
     {"name": "Haruna"},
       {"name": "Joseph"},
       {"name": "Damian"},
       {"name": "Isaac"},
       .   .   .],
  "links": [{"source": 0, "target": 2},
       {"source": 0, "target": 12},
       {"source": 0, "target": 20},
       {"source": 0, "target": 23},
       {"source": 0, "target": 31},
       .   .   .]}
```

Graph visualization with D3.js

D3.js provides us with the `d3.layout.force()` function, which uses the **Force Atlas** layout algorithm and helps us to visualize our graph. See `Chapter 3`, *Getting to Grips with Visualization*, for instructions of how to create D3.js visualizations.

First, we need to define the CSS style for the nodes, links, and node labels:

```
<style>

.link {
  fill: none;
  stroke: #666;
  stroke-width: 1.5px;
}

.node circle {
  fill: steelblue;
  stroke: #fff;
  stroke-width: 1.5px;
}

.node text {
  pointer-events: none;
  font: 10px sans-serif;
}
</style>
```

Then, we need to refer to the d3.js library:

```
<script src="http://d3js.org/d3.v3.min.js"></script>
```

Then, we need to define the `width` and `height` for the `svg` container and include it in the `body` tag:

```
var width = 1100,
    height = 800

var svg = d3.select("body").append("svg")
    .attr("width", width)
    .attr("height", height);
```

Now, we define the properties of the Force layout, such as `gravity`, `distance`, and `size`:

```
var force = d3.layout.force()
    .gravity(.05)
    .distance(150)
    .charge(-100)
    .size([width, height]);
```

Then, we need to acquire the data of the graph using the JSON format. We will configure the parameters for `nodes` and `links`:

```
d3.json("newJson.json", function(error, json) {
  force
      .nodes(json.nodes)
      .links(json.links)
      .start();
```

> For a complete reference on the D3.js Force Layout implementation, follow this link:
> `https://github.com/mbostock/d3/wiki/Force-Layout`

Then, we define the links as lines from the JSON data:

```
var link = svg.selectAll(".link")
    .data(json.links)
  .enter().append("line")
    .attr("class", "link");

var node = svg.selectAll(".node")
    .data(json.nodes)
  .enter().append("g")
    .attr("class", "node")
    .call(force.drag);
```

Now, we define the node as circles of size 6 and include the labels for each node:

```
node.append("circle")
    .attr("r", 6);

node.append("text")
      .attr("dx", 12)
      .attr("dy", ".35em")
      .text(function(d) { return d.name });
```

Finally, with the `tick` function, the Force layout simulation runs step by step:

```
force.on("tick", function() {
  link.attr("x1", function(d) { return d.source.x; })
      .attr("y1", function(d) { return d.source.y; })
      .attr("x2", function(d) { return d.target.x; })
      .attr("y2", function(d) { return d.target.y; });

    node.attr("transform", function(d) {
            return "translate(" + d.x + "," + d.y + ")"; })
    });
});
</script>
```

In the following screenshot, we can see the result of the visualization. In order to run the visualization, we just need to open a Command Terminal and run the following command:

```
>>python -m http.server 8000
```

Then, you just need to open a web browser and type in `http://localhost:8000/ForceGraph.html`. On the HTML page, we can see our Facebook graph with a gravity effect and we can interactively drag and drop the nodes:

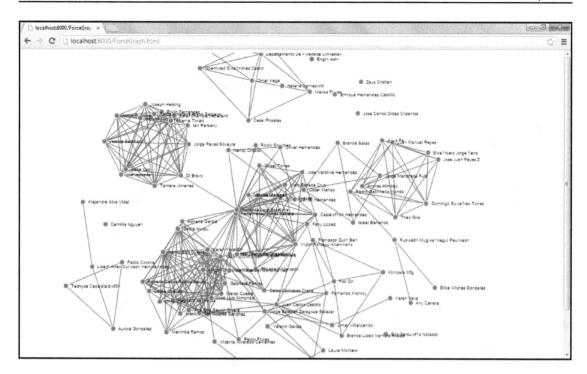

 All the code and datasets of this chapter may be found in the author's GitHub repository at:
`https://github.com/hmcuesta/PDA_Book/Chapter1`

The complete code for the visualization is listed here:

```
<meta charset="utf-8">
<style>

.link {
  fill: none;
  stroke: #666;
  stroke-width: 1.5px;
}
.node circle {
  fill: steelblue;
  stroke: #fff;
  stroke-width: 1.5px;
}

.node text {
  pointer-events: none;
```

```
        font: 10px sans-serif;
}
</style>
<body>
<script src="http://d3js.org/d3.v3.min.js"></script>
<script>

var width = 1100,
    height = 800

var svg = d3.select("body").append("svg")
    .attr("width", width)
    .attr("height", height);

var force = d3.layout.force()
    .gravity(.05)
    .distance(150)
    .charge(-100)
    .size([width, height]);

d3.json("newJson.json", function(error, json) {
  force
      .nodes(json.nodes)
      .links(json.links)
      .start();

  var link = svg.selectAll(".link")
      .data(json.links)
    .enter().append("line")
      .attr("class", "link");

  var node = svg.selectAll(".node")
      .data(json.nodes)
    .enter().append("g")
      .attr("class", "node")
      .call(force.drag);

  node.append("circle")
    .attr("r", 6);

  node.append("text")
      .attr("dx", 12)
      .attr("dy", ".35em")
      .text(function(d) { return d.name });

  force.on("tick", function() {
    link.attr("x1", function(d) { return d.source.x; })
        .attr("y1", function(d) { return d.source.y; })
```

```
        .attr("x2", function(d) { return d.target.x; })
        .attr("y2", function(d) { return d.target.y; });

    node.attr("transform", function(d) { return "translate(" + d.x + "," +
d.y + ")"; });
  });
});
</script>
</body>
```

Summary

In this chapter, we worked on how to obtain and visualize a Facebook graph, detect communities, use color nodes, and apply some layouts with Gephi, such as **Yifan Hu, Force Atlas**, and **Fruchterman-Reingold**. Then, we introduced some statistical methods to get aggregate information, such as degree, centrality, distribution, and ratio. Finally, we developed our own visualization tool with D3.js, transforming the data from `.gdf` into JSON.

In the next chapter, we will present a short introduction to the Twitter API to retrieve, visualize, and analyze tweets. Then, we will proceed to perform a sentiment analysis.

11
Working with Twitter Data

In this chapter, we will see how to acquire data from Twitter, exploring the fundamental elements of interaction such as retweets, likes, and trending topics. Initially, we introduce the **Twitter API** with Python. Then we will distinguish the basic elements of the Twython library and the use of **OAuth** as an authentication process. Finally, we will perform data acquisition in real time, applying data streaming.

In this chapter, we will cover:

- The anatomy of Twitter data
- Using OAuth to access the Twitter API
- Getting started with Twython
- Streaming API

With the proliferation of social network sites, we can see what people are talking about in real time and on a large scale. However, we need to be cautious because social networks tend to be noisy; that is why in this case we will need as much data as we can get in order to obtain a true representation of what people think.

Mining Twitter is one of the best ways to find out what people are talking about. Social media sites help us to communicate and to transfer knowledge in a very short format, in the case of Twitter, 140 characters.

The anatomy of Twitter data

Twitter is a Social Network site which provides a micro-blogging service for sharing text messages up to 140 characters long (Tweets). We can retrieve several types of data from Twitter, such as Tweets, Followers, Likes, Direct Messages, and Trending Topics.

We can create a new Twitter account with the following link:

```
http://twitter.com
```

Tweet

A Tweet is the name of the 140-character text message. However, we can get more information than the text message itself, such as date and time, links, user mentions (@), hash tags (#), re-tweet count, local language, favorites count, and geocode. In the following screenshot, we can observe a tweet re-tweeted **745** times, with 336 likes, a hashtag (**#WarEagle**), and user mentions (**@FootballAU** and **@CoachGuzMalzahn**):

Followers

Users on Twitter can follow other users, creating a directed graph (see Chapter 10, *Working with Social Graphs*) with a lot of possibilities for analysis, such as centrality and community clustering. In this case, the relationship is not mutual by default, so in Twitter we have two kinds of degrees in and out. This can be very useful when we want to find the most influent individual in a group, or which individual is in between different groups.

We can follow the Twitter engineering team's blog here:
`https://engineering.twitter.com/`

Trending topics

Twitter trends are words or #hashtags with a high popularity with Twitter users in a specific moment and place. Trending topics is a big area for data analysis, with topics such as how to detect trends and predict future trends; these are; main topics in information diffusion theory. In the following screenshot, we can see the dialog box to change tailored trends (trends based on your location and who you follow on Twitter) just by changing your location:

Using OAuth to access Twitter API

In order to have access to the Twitter API, we will use a **Token-Based Authentication** system. Twitter applications are required to use **OAuth**, which is an open standard for authorization. OAuth allows Twitter users to enter their username and password in order to obtain four strings (**Token**). The Token allows users to connect to the Twitter API without using their username and password. In this chapter, we will use the current version of Twitter REST API, 1.1, released on June 11, 2013, which established the use of OAuth authentication as mandatory to retrieve data from Twitter.

For more information about Token-Based Authentication systems, please refer to this link:
`http://bit.ly/bgbmnK`

First, we need to follow this link and sign in with our Twitter**username** and **password,** as shown in the following screenshot:

```
https://dev.twitter.com/apps
```

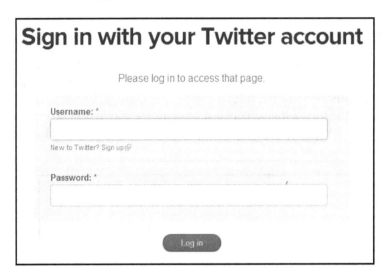

Then, we click on the **Create a new application** button (see following screenshot) and enter the application details:

- **Name**: PracticalDataAnalysisBook (it can be anything you like, but you cannot use the word *Twitter* in the name)
- **Description**: Practical Data Analysis Book Examples (can be anything you like)
- **Website**: Can be your personal blog or website
- **Callback URL**: Can be left blank

Next, we need to enter the CAPCHA and click on the **Create** button:

Finally, on the next details screen, we will click on **Create my access token** (sometimes you need to manually refresh the page after a few seconds). In the following screenshot, we can see our four strings for authentication: **Consumer key**, **Consumer secret**, **Access token**, and **Access token secret**:

OAuth settings

Your application's OAuth settings. Keep the "Consumer secret" a secret. This key should never be human-readable in your application.

Access level	Read-only About the application permission model
Consumer key	██████████████
Consumer secret	██████████████████████
Request token URL	`https://api.twitter.com/oauth/request_token`
Authorize URL	`https://api.twitter.com/oauth/authorize`
Access token URL	`https://api.twitter.com/oauth/access_token`
Callback URL	None
Sign in with Twitter	No

Your access token

Use the access token string as your `oauth_token` and the access token secret as your `oauth_token_secret` to sign requests with your own Twitter account. Do not share your `oauth_token_secret` with anyone.

Access token	████████████████████████
Access token secret	████████████████████████
Access level	Read-only

Now, we may use this access token with multiple user timelines on multiple websites through the Twitter Search API. However, this is restricted to 180 requests/queries per 15 minutes.

We can find more information about the Twitter **Search API**, limitations, best practices and rate limits at the following link:
`https://dev.twitter.com/docs/using-search`

Getting started with Twython

In this chapter, we will use**Twython 3**, which is a Python wrapper of the Twitter API v1.1. We can download the latest version of Twython from the **pypi** Python website:

```
https://pypi.python.org/pypi/twython
```

Then, we need to unzip and open the `twython` folder, and finally, we install the `twython` module with the following command:

```
>>> python3 setup.py install
```

Or, we can also install Twython through pip with the following command:

```
MacBook-Pro-de-Hector-3:~ hectorc$ sudo pip install twython
The directory '/Users/hectorc/Library/Caches/pip/http' or its parent directory is not owned by
 the current user and the cache has been disabled. Please check the permissions and owner of t
hat directory. If executing pip with sudo, you may want sudo's -H flag.
The directory '/Users/hectorc/Library/Caches/pip' or its parent directory is not owned by the
current user and caching wheels has been disabled. check the permissions and owner of that dir
ectory. If executing pip with sudo, you may want sudo's -H flag.
Collecting twython
  Downloading twython-3.4.0.tar.gz
Collecting requests>=2.1.0 (from twython)
  Downloading requests-2.10.0-py2.py3-none-any.whl (506kB)
    100% |████████████████████████████████| 512kB 316kB/s
Collecting requests_oauthlib>=0.4.0 (from twython)
  Downloading requests_oauthlib-0.6.1-py2.py3-none-any.whl
Collecting oauthlib>=0.6.2 (from requests_oauthlib>=0.4.0->twython)
  Downloading oauthlib-1.1.2.tar.gz (111kB)
    100% |████████████████████████████████| 112kB 380kB/s
Installing collected packages: requests, oauthlib, requests-oauthlib, twython
  Running setup.py install for oauthlib ... done
  Running setup.py install for twython ... done
Successfully installed oauthlib-1.1.2 requests-2.10.0 requests-oauthlib-0.6.1 twython-3.4.0
```

We can find the complete reference documentation for Twython in the following link:

```
https://twython.readthedocs.org/en/latest/index.html
```

> We can find a complete list of Twitter libraries for several programming languages, such as Java, C#, Python, and so on, in the following link:
> `https://dev.twitter.com/docs/twitter-libraries`

Simple search using Twython

In this example, we will perform a search of the word *python* and we will print the complete list of statuses in order to understand the format of the retrieved tweets.

First, we need to import the `Twython` object from the `twython` library:

```
from twython import Twython
```

Then, we need to define the four strings created with OAuth (see the *Using OAuth to access Twitter API* section):

```
ConsumerKey  = "..."
ConsumerSecret = "..."
AccessToken = "..."
AccessTokenSecret = "..."
```

Now, we need to instantiate the `Twython` object, giving the access token strings as the parameters:

```
twitter = Twython(ConsumerKey,
                  ConsumerSecret,
                  AccessToken,
                  AccessTokenSecret)
```

Next, we will perform the `search` with the search method, specifying the search query text in the keyword argument `q`:

```
result = twitter.search(q="python")
```

 Twython converts the JSON sent to us from Twitter to a naÃ¯ve Python object. However, if the authentication fails, the search will retrieve an error message like this:

```
{"errors":[{"message":"Bad Authentication data",
            "code":215}]}
```

Finally, we will iterate the `result["statuses"]` list and print each status (tweet):

```
for status in result["statuses"]:
    print(status)
```

The output of each status is retrieved in a JSON-like structure and will look like this:

```
{'contributors': None,
 'truncated': False,
 'text': 'La théorie du gender.... Genre Monty python !
http://t.co/3nTUhVR9Xm',
 'in_reply_to_status_id': None,
 'id': 355755364802764801,
 'favorite_count': 0,
 'source': '<a href="http://twitter.com/download/iphone"
rel="nofollow">Twitter for iPhone</a>',
 'retweeted': False,
 'coordinates': None,
 'entities': {'symbols': [],
        'user_mentions': [],
        'hashtags': [],
        'urls': [{'url': 'http://t.co/3nTUhVR9Xm',
            'indices': [46, 68],
            'expanded_url':
'http://m.youtube.com/watch?feature=youtube_gdata_player&v=ePCSA_N5QY0&desk
top_uri=%2Fwatch%3Fv%3DePCSA_N5QY0%26feature%3Dyoutube_gdata_player',
            'display_url': 'm.youtube.com/watch?feature=...'}]},
             'in_reply_to_screen_name': None,
        'in_reply_to_user_id': None,
        'retweet_count': 0,
        'id_str': '355755364802764801',
        'favorited': False,
        'user': {'follow_request_sent': False,
            'profile_use_background_image': True,
            'default_profile_image': False,
            'id': 1139268894,
            'verified': False,
            'profile_text_color': '333333',
            'profile_image_url_https':
'https://si0.twimg.com/profile_images/3777617741/d839f0d515c0997d8d18f55693
a4522c_normal.jpeg',
            'profile_sidebar_fill_color': 'DDEEF6',
            'entities': {'url': {'urls': [{'url':
'http://t.co/7ChRUG0D2Y',
                        'indices': [0, 22],
                        'expanded_url':
'http://www.manifpourtouslorraine.fr',
                        'display_url': 'manifpourtouslorraine.fr'}]},
                        'description': {'urls': []}}},
 'followers_count': 512,
 'profile_sidebar_border_color': 'C0DEED',
 'id_str': '1139268894',
 'profile_background_color': 'C0DEED',
```

```
 'listed_count': 9,
 'profile_background_image_url_https':
'https://si0.twimg.com/images/themes/theme1/bg.png',
 'utc_offset': None,
 'statuses_count': 249,
 'description': "ON NE LACHERA JAMAIS, RESISTANCE !!\r\nTous nÃ©s d'un
homme et d'une femme\r\nRetrait de la loi Taubira !\r\nRestons mobilisÃ©s
!",
 'friends_count': 152,
 'location': 'moselle',
 'profile_link_color': '0084B4',
 'profile_image_url':
'http://a0.twimg.com/profile_images/3777617741/d839f0d515c0997d8d18f55693a4
522c_normal.jpeg',
 'following': False,
 'geo_enabled': False,
 'profile_banner_url':
'https://pbs.twimg.com/profile_banners/1139268894/1361219172',
 'profile_background_image_url':
'http://a0.twimg.com/images/themes/theme1/bg.png',
 'screen_name': 'manifpourtous57',
 'lang': 'fr',
 'profile_background_tile': False,
 'favourites_count': 3,
 'name': 'ManifPourTous57',
 'notifications': False,
 'url': 'http://t.co/7ChRUG0D2Y',
 'created_at': 'Fri Feb 01 10:20:23 +0000 2013',
 'contributors_enabled': False,
 'time_zone': None,
 'protected': False,
 'default_profile': True,
 'is_translator': False},
 'geo': None,
 'in_reply_to_user_id_str': None,
 'possibly_sensitive': False,
 'lang': 'fr',
 'created_at': 'Fri Jul 12 18:27:43 +0000 2013',
 'in_reply_to_status_id_str': None,
 'place': None,
 'metadata': {'iso_language_code': 'fr', 'result_type': 'recent'}}
```

We can also restrict the results by navigating the structure of the JSON result. For example, to get only the `user` and `text` of the status, we can modify the `print` like this:

```
for status in result["statuses"]:
    print("user: {0} text: {1}".format(status["user"]["name"],
status["text"]))
```

The output of the first five statuses will look like this:

```
    user: RaspberryPi-Spy text: RT @RasPiTV: RPi.GPIO Basics Part 2, day 2
- Rev checking (Python &amp; Shell) http://t.co/We8PyOirqV
    user: Ryle Ploegsî?1 text: I really want the whole world to watch Monty
Python and the Holy Grail at least once. It's so freaking funny.
    user: Matt Stewart text: Casual Friday night at work... #snakes #scared
#python http://t.co/WVld2tVV8X
    user: Flannery O'Brien text: Kahn the Albino Burmese Python enjoying
the beautiful weather :) http://t.co/qvp6zXrG60
    user: Cian Clarke text: Estonia E-Voting Source Code Made Public
http://t.co/5wCulH4sht - open source, kind of! Python &amp; C
http://t.co/bo3CtukYoU
    . . .
```

Navigating through the JSON structure helps us to get only the information we need for our applications. We may pass multiple keyword arguments and also specify the result type with the `result_type="popular"` parameter.

> We can find a complete reference to get searches/tweets in the following link:
> `https://dev.twitter.com/docs/api/1.1/get/search/tweets`

Working with timelines

In this example, we will show how to retrieve our own timeline and a different user's timeline.

First, we need to import and instantiate the `Twython` object from the `twython` library:

```
from twython import Twython
ConsumerKey        = "..."
ConsumerSecret     = "..."
AccessToken        = "..."
AccessTokenSecret  = "..."
twitter = Twython(ConsumerKey,
                  ConsumerSecret,
                  AccessToken,
                  AccessTokenSecret)
```

Now, to get our own timeline, we will use the `get_home_timeline` method:

```
timeline = twitter.get_home_timeline()
```

Finally, we will iterate the timeline and print `user name`, `created_at`, and `text`:

```
for tweet in timeline:
    print(" User: {0} \n Created: {1} \n Text: {2} "
        .format(tweet["user"]["name"],
                tweet["created_at"],
                tweet["text"]))
```

The first five results of the code will look like this:

```
User: Ashley Mayer
Created: Fri Jul 12 19:42:46 +0000 2013
Text: Is it too late to become an astronaut?
User: Yves Mulkers
Created: Fri Jul 12 19:42:11 +0000 2013
Text: The State of Pharma Market Intelligence http://t.co/v0f1DH7KZB
User: Olivier Grisel
Created: Fri Jul 12 19:41:53 +0000 2013
Text: RT @stanfordnlp: Deep Learning Inside: Stanford parser quality
improved with new CVG model. Try the englishRNN.ser.gz model.
http://t.co/jE...
user: Stanford Engineering
Created: Fri Jul 12 19:41:49 +0000 2013
Text: Ralph Merkle (U.C. Berkeley), Martin Hellman (#Stanford
Electrical #Engineering) and Whitfield Diffie... http://t.co/4y7Gluxu8E
User: Emily C Griffiths
Created: Fri Jul 12 19:40:45 +0000 2013
Text: What role for equipoise in global health? Interesting Lancet
blog: http://t.co/2FA6ICfyZX
    . . .
```

On the other hand, if we want to retrieve a specific user's timeline, such as `stanfordeng`, we will use the `get_user_timeline` method with the `screen_name` parameter to define the selected user, and we can also restrict the number of results to five with the `count` parameter:

```
tl = twitter.get_user_timeline(screen_name = "stanfordeng",
                               count = 5)
for tweet in tl:
    print(" User: {0} \n Created: {1} \n Text: {2} "
        .format(tweet["user"]["name"],
                tweet["created_at"],
                tweet["text"]))
```

The first five statuses of the Stanford Engineering (@stanfordeng) timeline will look like this:

```
User: Stanford Engineering
Created: Fri Jul 12 19:41:49 +0000 2013
Text: Ralph Merkle (U.C. Berkeley), Martin Hellman (#Stanford
Electrical #Engineering) and Whitfield Diffie... http://t.co/4y7Gluxu8E
User: Stanford Engineering
Created: Fri Jul 12 15:49:25 +0000 2013
Text: @nitrogram W00t!! ;-)
User: Stanford Engineering
Created: Fri Jul 12 15:13:00 +0000 2013
Text: Stanford team (@SUSolarCar) to send newest creation, solar car
#luminos for race in Australia: http://t.co/H5bTSEZcYS. via @paloaltoweekly
User: Stanford Engineering
Created: Fri Jul 12 02:50:00 +0000 2013
Text: Congrats! MT @coursera: Coursera closes w 43M in Series B.
Doubling in size to focus on mobile, apps platform &amp; more!
http://t.co/WTqZ7lbBhd
User: Stanford Engineering
Created: Fri Jul 12 00:57:00 +0000 2013
Text: Engineers can really benefit from people who can make intuitive
or creative leaps. ~Stanford Electrical Engineering Prof. My Le  #quote
```

We can find a complete reference for the `home_timeline` and `user_timeline` methods in the following links:
`http://bit.ly/nEpIW9`
`http://bit.ly/QpgvRQ`

Working with followers

In this example, we will show how to retrieve the list of followers of specific Twitter users.

First, we need to import and instantiate the `Twython` object from the `twython` library:

```
from twython import Twython
ConsumerKey       = "..."
ConsumerSecret    = "..."
AccessToken       = "..."
AccessTokenSecret = "..."
twitter = Twython(ConsumerKey,
                  ConsumerSecret,
                  AccessToken,
                  AccessTokenSecret)
```

Next, we will return the list of followers with the `get_followers_list` method, using `screen_name` (username) or `user_id` (Twitter user ID):

```
followers = twitter.get_followers_list(screen_name="hmcuesta")
```

Next, we iterate over the followers`["users"]` list and print all the followers:

```
for follower in followers["users"]:
    print(" {0} \n ".format(follower))
```

Each user will look like this:

```
{'follow_request_sent': False,
 'profile_use_background_image': True,
 'default_profile_image': False,
 'id': 67729744,
 'verified': False,
 'profile_text_color': '333333',
 'profile_image_url_https':
'https://si0.twimg.com/profile_images/374723524/iconD_normal.gif',
 'profile_sidebar_fill_color': 'DDEEF6',
 'entities': {'description': {'urls': []}},
 'followers_count': 7,
 'profile_sidebar_border_color': 'C0DEED',
 'id_str': '67729744',
 'profile_background_color': 'C0DEED',
 'listed_count': 0,
 'profile_background_image_url_https':
'https://si0.twimg.com/images/themes/theme1/bg.png',
 'utc_offset': -21600,
 'statuses_count': 140,
 'description': '',
 'friends_count': 12,
 'location': '',
 'profile_link_color': '0084B4',
 'profile_image_url':
'http://a0.twimg.com/profile_images/374723524/iconD_normal.gif',
 'following': False,
 'geo_enabled': False,
 'profile_background_image_url':
'http://a0.twimg.com/images/themes/theme1/bg.png',
 'screen_name': 'jacobcastelao',
 'lang': 'en',
 'profile_background_tile': False,
 'favourites_count': 1,
 'name': 'Jacob Castelao',
 'notifications': False,
 'url': None,
```

```
'created_at': 'Fri Aug 21 21:53:01 +0000 2009',
'contributors_enabled': False,
'time_zone': 'Central Time (US & Canada)',
'protected': True,
'default_profile': True,
'is_translator': False}
```

Finally, we will print only the user (screen_name), name, and number of tweets (statuses_count):

```
for follower in followers["users"]:
    print(" user: {0} \n name: {1} \n Number of tweets: {2} \n"
        .format(follower["screen_name"],
                follower["name"],
                follower["statuses_count"]))
```

For the first five followers, the results of the preceding code will look like this:

```
user: katychuang
name: Kat Chuang, PhD
number of tweets: 1991
user: fractalLabs
name: Fractal Labs
number of tweets: 105
user: roger_yau
name: roger yau
number of tweets: 70
user: DataWL
name: Data Without Limits
number of tweets: 1168
user: abhi9u
name: Abhinav Upadhyay
number of tweets: 5407
```

We can find a complete reference for the get_followers_list method in the following link:
https://dev.twitter.com/docs/api/1.1/get/followers/list

Working with places and trends

In this example, we will retrieve the trending topics closest to a specific location. In order to specify the location, the Twitter API uses the Yahoo! **WOEID (Where On Earth ID)**.

First, we need to import and instantiate the `Twython` object from the `twython` library:

```
from twython import Twython
ConsumerKey       = "..."
ConsumerSecret    = "..."
AccessToken       = "..."
AccessTokenSecret = "..."
twitter = Twython(ConsumerKey,
                  ConsumerSecret,
                  AccessToken,
                  AccessTokenSecret)
```

Next, we will use `get_place_trends` and we define the place with the `id = (WOEID)` parameter:

```
result = twitter.get_place_trends(id = 23424977)
```

We can find a complete reference for the `get_place_trends` method in the following link:
`https://dev.twitter.com/docs/api/1.1/get/trends/closest`

The easiest way to get the WOEID is through the console of **Yahoo! Query Language (YQL)**, which uses a SQL-like syntax, so if we want to find the WOEID of `"Denton, TX"` the string query will look like this:

```
select * from geo.places where text="Denton, TX"
```

We can find the console in the following link and we can test the string query just by clicking on the **Test** button:

`http://developer.yahoo.com/yql/console/`

In the following screenshot, we can see the result of the query in JSON format, and an arrow points to the WOEID attribute:

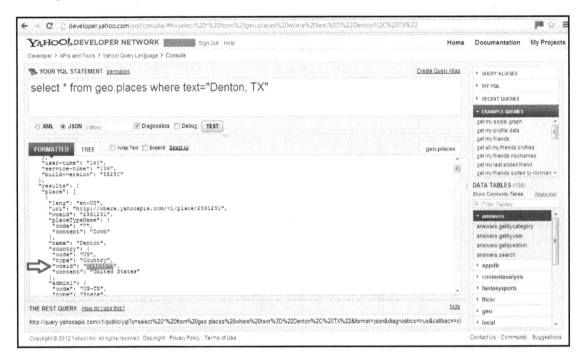

Finally, we will iterate the `result` and print the `name` of each `trend`:

```
if result:
    for trend in result[0].get("trends", []):
        print("{0} \n".format(trend["name"]))
```

The trending topics in `"Denton, TX"` will look like this:

```
#20FactsAboutMyBrother
#ImTeamTwist
Ho Lee Fuk
#FamousTamponQuotes
#TopTenHoeQuotes
#BaeLiterature
Cosart
KTVU
NTSB
Pacific Rim
```

Working with user data

In this example, we will retrieve information about a specific user, such as number of tweets, followers, following profile image, and so on.

First, we need to import and instantiate the `Twython` object from the `twython` library, then we just need to look up the user using the screen name of the user, as shown in this statement:

```
users = twitter.lookup_user(screen_name = "datacampmx")
```

Finally, we just need to print the information of the user; this is very useful, for example, if we need to get information about our followers and detect who is the top followed, or to get their profile images. We may see the complete code in the following screenshot:

```
twitter = Twython(ConsumerKey,
                  ConsumerSecret,
                  AccessToken,
                  AccessTokenSecret)
users = twitter.lookup_user(screen_name = "datacampmx")
for user in users:
    print "image: "      + user["profile_image_url_https"]
    print "twitts: "     + str(user["statuses_count"])
    print "followers: "  + str(user["followers_count"])
    print "followings: " + str(user["friends_count"])
    print "time zone: "  + user["time_zone"]

image: https://pbs.twimg.com/profile_images/744959798962094080/edTbc-mg_normal.jpg
twitts: 361
followers: 266
followings: 669
time zone: Pacific Time (US & Canada)
```

Streaming API

The **Streaming API** generates a persistent connection through HTTP; this means that we may keep a query running while waiting for a response, and whenever new information is available, the server sends the response to the client. The use of the Streaming API reduces network latency by significantly reducing the number of connections to the server.

Social network sites are an excellent source of streaming data due to their user-generated content, updated each second. For example, when we talk about an event like the Olympic Games or Wall Street options day trading, we need to know the status of certain events in real time. When you need to process data in real time, you may use the Streaming API. Twython provides us with streaming capabilities; we may run a filter command.

First, we need to import and instantiate the `TwythonStreamer` object from the `twython` library:

```
from twython import TwythonStreamer

ConsumerKey   = "..."
ConsumerSecret = "..."
AccessToken = "..."
AccessTokenSecret = "..."
```

Next, we will create a class named `MyStreamer`; this class will help us to define the streaming object. In order to achieve this, we will define two methods. The first one is `on_success`, which will define the logic of the streaming; for this example, we are just going to print the text of the tweets. The second is an `on_error` method to print a log of errors when they happen. This example will keep running until we kill the process; if we want to stop the streaming from the code, we must include the `self.disconnect()` statement in the `on_success` method:

```
class MyStreamer(TwythonStreamer):
    def on_success(self, data):
        if 'text' in data:
            print data['text'].encode('utf-8')

    def on_error(self, status_code, data):
        print status_code, data
```

Then, we just initiate the `stream` object by calling the `MyStreamer` constructor, and we give it our OAuth keys as a parameter:

```
stream = MyStreamer(ConsumerKey,
                    ConsumerSecret,
                    AccessToken,
                    AccessTokenSecret)
```

Finally, we will invoke the `statuses.filter` method from the stream object and we will use the `track` parameter to define the keyword of tweets we will consume:

```
stream.statuses.filter(track='cnn')
```

We can see the result of the code execution in the following screenshot:

```
In [*]:  from twython import TwythonStreamer

         ConsumerKey  = "41E1twaPLNgsn0Q4VS5g"
         ConsumerSecret = "augwZxzQGsJuyfzLzGn0ASpherv2YgpeLTKEXXFk"
         AccessToken = "141340589-WhKonOAcDmCX1MVJNpd3UEB2gvzZt2nmPBJfMy3o"
         AccessTokenSecret = "LDsdOa9Mex2yZ0AMud4eUe0mlcqvvsycwp0yneSWQw"

         class MyStreamer(TwythonStreamer):
             def on_success(self, data):
                 if 'text' in data:
                     print data['text'].encode('utf-8')

             def on_error(self, status_code, data):
                 print status_code, data

         stream = MyStreamer(ConsumerKey,
                             ConsumerSecret,
                             AccessToken,
                             AccessTokenSecret)

         stream.statuses.filter(track='cnn')
```

```
"Frustrated"??? More like mad. Angry. Righteous indignation. "Frustrated" is a traffic jam. This was a man's LIFE!!!
https://t.co/Wuwc6vI5OL
RT @ssirah: Capek2 ngitung ulang ktp 😀😀 https://t.co/K7JoTPyTCC
RT @magnifier661: 👉CNN OFFICIAL BLACKOUT👋
JULY 1ST - JULY14TH
@Toyota @etrade @WellsFargo @sprint @GEICO
#BlackOutCNN #Trump2016 https://t…
#USA#news Wimbledon 2016: Dresses too revealing?: Nike's flimsy attire for women's pros at Wimbledon is gener... http
s://t.co/9ZyME2XP8L
الدا حنية المصرية تعلن مقتل كاهن كنيسة مارجرجس في #العريش برصاص مجهول.. و" #داعش" يتبنى :RT @cnnarabic
https://t.co/sJQYHH4hAE
مصر  #الكنيسة#
RT @CNNPolitics: .@realDonaldTrump: "I totally disavow the Ku Klux Klan" https://t.co/PE1ISCG7gC https://t.co/DImhdKS
W6B
#DOMA# Wimbledon 2016: Dresses too revealing? https://t.co/9ZyME2XP8L
Feds: Stop driving these Honda models right now https://t.co/3caFJLe1lf
Wimbledon 2016: Dresses too revealing? https://t.co/H3wCm47XIX
RT @CNN: President Obama on #Istanbul terror attack: "The prayers of the American people are with the people of Turke
y" https://t.co/dnDQyf…
@Brainiac_13 @CNN I am no more IPhones for me I've grown tired of them
```

Summary

In this chapter, we covered the basic functions of the Twitter API, from signing in with OAuth to location trends, how to perform simple queries, and how to get data from a specific Twitter user. Then we introduced the concept of streaming data using the Streaming API provided by the `Twython` library.

In the next chapter, we will present the basic concepts of **MongoDB** and how we can perform aggregation queries with a large amount of data.

12
Data Processing and Aggregation with MongoDB

Aggregation queries are a very common way to get summarized data by counting or adding features onto our dataset. **MongoDB** provides us with different ways to get aggregated data fast and easy. In this chapter, we will explore the basic features of MongoDB as well as two ways to get summarized data using the `Group` function and the **Aggregation** framework.

In this chapter, we will cover the following topics:

- Getting started with MongoDB
- Data processing
- Aggregation framework

In Chapter 2, *Preprocessing the Data*, we introduced **NoSQL (Not Only SQL)** databases and their types (document-based, graph-based, and key-value stores). NoSQL databases provide key advantages to the user, such as scalability, high availability, and processing speed. Due to the distributed nature of the NoSQL technology, if we want to scale a NoSQL database, we just need to add machines to the cluster to meet demand (horizontal scaling). Most NoSQL databases are open source (such as **MongoDB**, **Cassandra**, and **HBase**); this means that we can download, implement, and scale them at a very low cost.

Getting started with MongoDB

MongoDB is a very popular document-oriented NoSQL database. It provides a high performance engine for storage and query retrieval. In a document-oriented database, we store the data into the collections of documents; in this case, it is JSON-like documents called **BSON** (**Binary JSON**), which provide us with a dynamic schema data structure. MongoDB implements functionalities such as ad hoc queries, replication, load balancing, Aggregation, and Map-Reduce. MongoDB is perfect for an operational database. However, its capabilities as a transactional data source are limited. We can see in the following image the similarities between the structures of a relational database (RDBMS) and MongoDB. We may find more information about MongoDB from its official website:

```
http://www.mongodb.org/
```

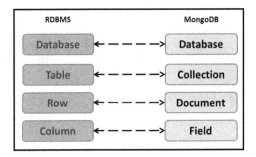

For a complete reference about SQL databases, refer to this URL:
```
http://www.w3schools.com/sql/
```

From the preceding image, we can see that the internal structure of MongoDB is very similar to a relational database. However, in this case, in our database, we have a set of collections with BSON documents in it, without any previous schema defined. And not all documents in a collection must have the same schema. The installation in iOS is very easy using **Homebrew**, which is a free and open source software package management system that simplifies the installation of software. We use the `brew install` command and the name of the software, as is shown here:

```
iMac-de-Hector:~ hectorcuesta1$ brew install mongodb
==> Downloading https://homebrew.bintray.com/bottles/mongodb-3.2.4.yosemite.bott
################################################################## 100.0%
==> Pouring mongodb-3.2.4.yosemite.bottle.tar.gz
==> Caveats
To have launchd start mongodb at login:
    ln -sfv /usr/local/opt/mongodb/*.plist ~/Library/LaunchAgents
Then to load mongodb now:
    launchctl load ~/Library/LaunchAgents/homebrew.mxcl.mongodb.plist
Or, if you don't want/need launchctl, you can just run:
    mongod --config /usr/local/etc/mongod.conf
==> Summary
🍺 /usr/local/Cellar/mongodb/3.2.4: 17 files, 208.7M
```

Once we install the MongoDB engine, we will run the `mongod` command from Terminal. This will show that the engine is ready to listen for new connections, as shown in the following screenshot.

 You must define your data folder (by default, this is located in the root directory, ex., called /data/db) before the execution of the engine.

```
iMac-de-Hector:~ hectorcuesta1$ mongod
2016-04-13T09:45:43.025-0500 I CONTROL  [initandlisten] MongoDB starting : pid=5541 port=27017 dbpath=/data/db
64-bit host=iMac-de-Hector.local
2016-04-13T09:45:43.026-0500 I CONTROL  [initandlisten] db version v3.2.4
2016-04-13T09:45:43.026-0500 I CONTROL  [initandlisten] git version: e2ee9ffcf9f5a94fad76802e28cc978718bb7a30
2016-04-13T09:45:43.026-0500 I CONTROL  [initandlisten] allocator: system
2016-04-13T09:45:43.026-0500 I CONTROL  [initandlisten] modules: none
2016-04-13T09:45:43.026-0500 I CONTROL  [initandlisten] build environment:
2016-04-13T09:45:43.026-0500 I CONTROL  [initandlisten]     distarch: x86_64
2016-04-13T09:45:43.026-0500 I CONTROL  [initandlisten]     target_arch: x86_64
2016-04-13T09:45:43.026-0500 I CONTROL  [initandlisten] options: {}
2016-04-13T09:45:43.026-0500 I STORAGE  [initandlisten] exception in initAndListen: 29 Data directory /data/db
not found., terminating
2016-04-13T09:45:43.026-0500 I CONTROL  [initandlisten] dbexit:  rc: 100
iMac-de-Hector:~ hectorcuesta1$ mongod
2016-04-13T09:46:59.909-0500 I CONTROL  [initandlisten] MongoDB starting : pid=5550 port=27017 dbpath=/data/db
64-bit host=iMac-de-Hector.local
2016-04-13T09:46:59.910-0500 I CONTROL  [initandlisten] db version v3.2.4
2016-04-13T09:46:59.910-0500 I CONTROL  [initandlisten] git version: e2ee9ffcf9f5a94fad76802e28cc978718bb7a30
2016-04-13T09:46:59.910-0500 I CONTROL  [initandlisten] allocator: system
2016-04-13T09:46:59.910-0500 I CONTROL  [initandlisten] modules: none
2016-04-13T09:46:59.910-0500 I CONTROL  [initandlisten] build environment:
2016-04-13T09:46:59.910-0500 I CONTROL  [initandlisten]     distarch: x86_64
2016-04-13T09:46:59.910-0500 I CONTROL  [initandlisten]     target_arch: x86_64
2016-04-13T09:46:59.910-0500 I CONTROL  [initandlisten] options: {}
2016-04-13T09:46:59.910-0500 I STORAGE  [initandlisten] wiredtiger_open config: create,cache_size=4G,session_ma
x=20000,eviction=(threads_max=4),config_base=false,statistics=(fast),log=(enabled=true,archive=true,path=journa
l,compressor=snappy),file_manager=(close_idle_time=100000),checkpoint=(wait=60,log_size=2GB),statistics_log=(wa
it=0),
2016-04-13T09:47:02.519-0500 I FTDC     [initandlisten] Initializing full-time diagnostic data capture with dir
ectory '/data/db/diagnostic.data'
2016-04-13T09:47:02.519-0500 I NETWORK  [HostnameCanonicalizationWorker] Starting hostname canonicalization wor
ker
2016-04-13T09:47:04.408-0500 I NETWORK  [initandlisten] waiting for connections on port 27017
2016-04-13T09:50:22.606-0500 I NETWORK  [initandlisten] connection accepted from 127.0.0.1:62468 #1 (1 connecti
on now open)
2016-04-13T09:50:22.612-0500 I NETWORK  [initandlisten] connection accepted from 127.0.0.1:62469 #2 (2 connecti
ons now open)
```

Database

In MongoDB, the database is a physical container for our collections. Each database will create a set of files on the file system. In MongoDB, a database is created automatically on the fly when we save for the first time a document into a collection. However, administration tools, such as **Robomongo**, allow us to create databases. First, we need to create a connection from the Robomongo to MongoDB engine. We will provide a name in the **Name** field for the connection and the **Address** (IP Host and Port), then we click on **Save**. Now we can see a new **Connection**. Once we are connected with MongoDB, we will click on the name of the engine and select the **Create Database** option, as shown here:

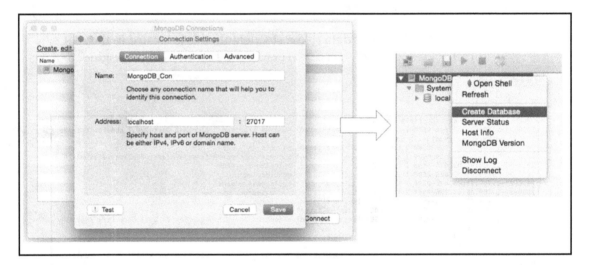

 You can find more information about it at:
https://robomongo.org/

Additionally, we can see the available databases with the show dbs command and the result can be seen in the terminal, as is shown in this screenshot:

Collection

A collection is a group of documents. MongoDB will create the collection implicitly as is done with the database. As MongoDB uses a schema-less model, we must specify the database and the collection where it will be stored. MongoDB provides a JavaScript function called `db.createCollection()` to create a collection manually, and we can also create collections from the Robomongo interface.

We can specify the database with the use `<database name>` command in the **Mongo Shell** and we can see the available collections in the database with the `show collections` command, as shown here:

```
> use Corpus
switched to db Corpus
> show collections
system.indexes
tweets
>
```

Collections can be split to distribute the collections documents across the MongoDB instances (**shards**). This process is called **sharding** and allows a horizontal scaling.

> We may find considerations about data modeling in MongoDB at
> `http://docs.mongodb.org/manual/core/data-modeling/`.

Document

A document is a record in MongoDB and implements a schema-less model. This means that the documents are not enforced to have the same set of fields or structure. However, in practice, the documents share a basic structure in order to perform queries and complex searches.

MongoDB uses a document format like **JSON (JavaScript Object Notation)**, stored in a binary representation called BSON. For Python programmers, we will use the same dictionary structure to represent the JSON format seen in Chapter 2, *Preprocessing the Data*. We may find the complete BSON specification at `http://bsonspec.org/`.

MongoDB uses a dot notation (`.`) to navigate through the JSON structure to access a field in the document or subdocument, for example, `<subdocument>. <field>`.

Mongo shell

Mongo shell is an interactive JavaScript console for MongoDB. Mongo shell comes as a standard feature in the MongoDB. We also have the option to try a small version of Mongo shell in the official website (see the following screenshot) what is good enough to start in MongoDB:

We can find the FAQ of Mongo shell at
`http://docs.mongodb.org/manual/faq/mongo/`.

Insert/Update/Delete

Now, we will explore the basic operations with MongoDB, and we will compare them with the analogous instructions in SQL just as references. If you already have some experience with relational database SQL language, this is going to be a natural transition:

- `INSERT` method in SQL:

```
INSERT INTO Collection (First_Name, Last_Name)
          Values ('Jan', 'Smith');
```

- `Insert` method in MongoDB:

```
db.collection.insert({ name: { first: 'Jan', last: 'Smith' } )
```

- UPDATE method in SQL:

```
UPDATE Collection
SET First_Name = 'Joan'
WHERE First_Name = 'Jan';
```

- `Update` method in MongoDB:

```
db.collection.update(
    { 'name.first': 'Jan' },
    { $set: { 'name.first': 'Joan' } }
)
```

- DELETE method in SQL:

```
DELETE FROM Collection
WHERE First_Name = 'Jan';
```

- `Delete` method in MongoDB:

```
db.collection.remove( { 'name.first' : 'Jan' }, safe=True )
```

> We can find the documentation for Core MongoDB Operations at
> `http://docs.mongodb.org/manual/crud/`.

Queries

In MongoDB, we can perform searches and retrieve data with two methods, `find` and `findOne`, and both are explained here:

- Selecting all elements in the Table Collection in SQL:

```
SELECT * FROM Collection
```

- Selecting all elements in the collection in MongoDB:

```
db.collection.find()
```

In the following screenshot, we can see the result of the find method in the Mongo shell:

```
> db.test.data.find()
{ "_id" : ObjectId("51eedee2d341516bbfdbc6ff"), "name" : { "first" : "Jan", "las
t" : "Smith" } }
{ "_id" : ObjectId("51eedf0cd341516bbfdbc700"), "name" : { "first" : "Damian", "
last" : "Cuesta" } }
{ "_id" : ObjectId("51eedf17d341516bbfdbc701"), "name" : { "first" : "Isaac", "l
ast" : "Cuesta" } }
> _
```

- Getting the number of documents retrieved in a query with SQL:

```
SELECT count(*) FROM Collection
```

- Getting the number of documents retrieved in a query with MongoDB:

```
db.collection.find().count()
```

- Query with specific criteria with SQL:

```
SELECT * FROM Collection
WHERE Last_Name = "Cuesta"
```

- Query with specific criteria with MongoDB:

```
db.collection.find({"name.last":"Cuesta"})
```

In the following screenshot, we can see the result of the find method using specific criteria in the Mongo shell:

```
> db.test.data.find({"name.last":"Cuesta"})
{ "_id" : ObjectId("51eedf0cd341516bbfdbc700"), "name" : { "first" : "Damian", "
last" : "Cuesta" } }
{ "_id" : ObjectId("51eedf17d341516bbfdbc701"), "name" : { "first" : "Isaac", "l
ast" : "Cuesta" } }
>
```

The findOne method retrieves a single document from a collection and does not return a list of documents (cursor). In the following screenshot, we can see the result of the findOne method in the Mongo shell:

```
> db.test.data.findOne()
{
        "_id" : ObjectId("51eedee2d341516bbfdbc6ff"),
        "name" : {
                "first" : "Jan",
                "last" : "Smith"
        }
}
>
```

We can find documentation for Read Operations at
`http://docs.mongodb.org/manual/core/read-operations/`.

When we want to test the query operation and the timing of the query, we will use the `explain` method. In the next screenshot, we can see the result of the `explain` method to find the efficiency of the queries and index use.

In the code listed here, we can see the use of the `explain` method in the `find` method:

```
db.collection.find({"name.last":"Cuesta"}).explain()
```

```
> db.test.data.find({"name.last":"Cuesta"}).explain()
{
        "cursor" : "BasicCursor",
        "isMultiKey" : false,
        "n" : 2,
        "nscannedObjects" : 3,
        "nscanned" : 3,
        "nscannedObjectsAllPlans" : 3,
        "nscannedAllPlans" : 3,
        "scanAndOrder" : false,
        "indexOnly" : false,
        "nYields" : 0,
        "nChunkSkips" : 0,
        "millis" : 0,
        "indexBounds" : {

        },
        "server" : "Hadoop-PC:27017"
}
```

Data preparation

In `Chapter 11`, *Working with Twitter Data*, we explored how to create a bag of words from the Tweets**Sentiment140** dataset. In this chapter, we will complement the example using MongoDB. First, we will prepare and transform the dataset from CSV into a JSON format in order to add it into a MongoDB collection.

We can download the Sentiment140 training and test data at
`http://help.sentiment14.com/for-students`.

We will download and open the test data; the columns represent `sentiment`, `id`, `date`, and via, user, and text. The first five records will look similar to this:

```
4,1,Mon May 11 03:21:41 UTC 2009,kindle2,yamarama,@mikefish  Fair
enough. But i have the Kindle2 and I think it's perfect   :)
4,2,Mon May 11 03:26:10 UTC 2009, jquery,dcostalis,Jquery is my new
best friend.
4,3,Mon May 11 03:27:15 UTC 2009,twitter,PJ_King,Loves twitter
4,4,Mon May 11 03:29:20 UTC 2009,obama,mandanicole,how can you not love
Obama? he makes jokes about himself.
4,5,Mon May 11 05:22:12 UTC 2009,lebron,peterlikewhat,lebron and
zydrunas are such an awesome duo
```

The first problem that we can see is that the field text includes the comma (`,`) character in it. This will be a problem if we want to read the file from Python. In order to solve this problem, we will perform data preparation in **OpenRefine** before we start working with the file. See `Chapter 2`, *Preprocessing the Data* for an introduction to OpenRefine.

Data transformation with OpenRefine

First, we need to run OpenRefine and import the `testdata manual 2009 06 14.csv` file. Then, we will select the number of columns (separate by commas), and click on the **create the project** button. In the following screenshot, we can see the interface of OpenRefine with the six columns, and we can rename the columns by clicking in each column, then selecting **Edit column/Rename this column**:

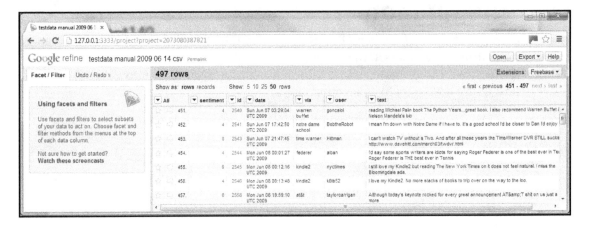

In order to delete the comma character from the field text, we need to click in the column text, and then select **Edit Cells/ Transform…**. Now, in the **Custom text transform on column text** window, we will use the `replace` function to eliminate all the commas from text, as shown in the next screenshot.

In the command listed here, we can see the function `replace` from **OpenRefine Expression Language** (**GREL**):

```
value.replace(",", "")
```

GREL implements a large selection of functions for **Strings**, **Arrays**, **Math**, **Dates**, and **Boolean**. We can find more information about it at the following URL:

```
https://github.com/OpenRefine/OpenRefine/wiki/GREL-Functions
```

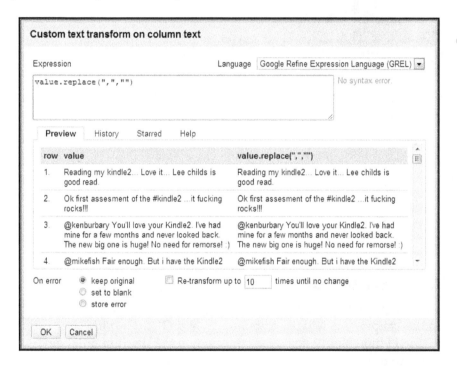

Finally, to export the dataset into a JSON format, we will select the **Export** select box, and then **Templating**. Then, we can see the **Templating Export** window (see the next screenshot) where we can define the final structure and row template in JSON format. Finally, we need to click on the **Export** button to download the test.json file into a file system location:

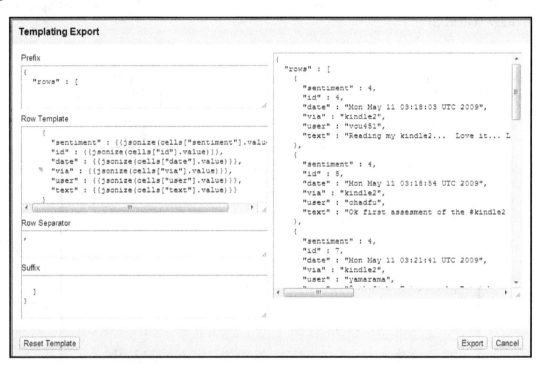

Inserting documents with PyMongo

With the dataset in JSON format, it will be much easier to insert records into the MongoDB collection. In this chapter, we will use Robomongo as a **GUI** (**Graphic User Interface**) tool and the pymongo Python moduleto install pymongo on the **IPython notebook** that we just need to use the !pip install command, as shown here:

```
In [1]: !pip install pymongo

        Collecting pymongo
          Downloading pymongo-3.2.2-cp27-none-macosx_10_10_intel.whl (262kB)
            100% |████████████████████████████████| 266kB 780kB/s
        Installing collected packages: pymongo
        Successfully installed pymongo-3.2.2

In [2]: import pymongo
```

Once we have pymongo installed, we can import the MongoClient object to create a connection with the MongoDB engine, as follows:

```
import json
from pymongo import MongoClient
```

> We can find complete documentation about PyMongo at
> http://api.mongodb.org/python/current/.

Next, we need to establish connection with pymongo using the Connection function:

```
con = MongoClient()
```

Then, we will select the Corpus database:

```
db = con.Corpus
```

Now, we will select the tweets collection where all our documents will be stored:

```
tweets = db.tweets
```

Finally, we will open the test.txt file as the structure of dictionaries with the json.load function:

```
with open("test.txt") as f:
    data = json.loads(f.read())
```

Then, we will iterate all the rows and insert them into the tweets collection:

```
for tweet in data["rows"]:
    tweets.insert(tweet)
```

If we prefer to use a graphic interface, we will establish connection in Robomongo, and we will create a new database as we see on the Robomongo interface, as seen here:

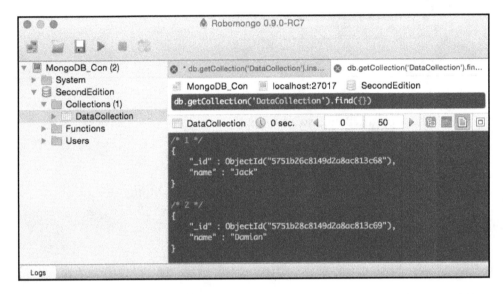

The result can be seen in Robomongo by navigating through the `Corpus` database and the `tweets` collection. Then, we click on the **find** option to retrieve all the documents in the collection.

All the code and datasets of this chapter can be found in the author's GitHub repository at
`https://github.com/hmcuesta/PDA_Book/tree/master/Chapter12`.

The complete code is here:

```
import json
from pymongo import MongoClient
con = MongoClient()
db = con.Corpus
tweets = db.tweets
with open("test.txt") as f:
    data = json.loads(f.read())
    for tweet in data["rows"]:
        tweets.insert(tweet)
```

Group

Aggregation function is a type of function used in data processing by grouping values into categories in order to find a significant meaning. The common aggregate functions include **Count**, **Average**, **Maximum**, **Minimum**, and **Sum**. However, we may perform more complicated statistical functions, such as **Mode** or **Standard Deviation**. Typically, the grouping is performed with the SQL Group by statement, as is shown in the following code; additionally, we may use aggregation functions, such as COUNT, MAX, MIN, and SUM in order to retrieve summarized information:

```
SELECT sentiment, COUNT(*)
FROM Tweets
GROUP BY sentiment
```

In MongoDB, we can use the Group function, which is similar to the SQL Group By statement. However, the Group function doesn't work in shared systems, and the result size is limited to 10,000 documents (20,000 in version 2.2 or higher). Due to this, the Group function is not very often used. Nevertheless, it is an easy way to find aggregate information when we have only one MongoDB instance.

In the following code, we can see the group function applied to the collection. The Group command needs a key, which is a field or fields to group. Then, we will define a reduce function, which will implement the aggregation function, in this case, the count of documents grouped by the sentiment field. Finally, we define the initial value for the aggregation result document:

```
db.collection.group({
    key:{sentiment:true},
    reduce: function(obj,prev{prev. sentimentsum += obj.c}),
    initial: {sentimentsum: 0}
});
```

> We can find the documentation of the function Group at
> http://bit.ly/15iICc5.

In the following code, we can see how to perform a grouping in PyMongo using the group function of the tweets collection:

```
from pymongo import MongoClient
con = MongoClient()
db = con.Corpus
tweets = db.tweets
```

```
categories = tweets.group(key={"sentiment":1},
                          condition={},
                          initial={"count": 0},
                          reduce="function(obj,    prev)
                                    {prev.count++;}")
for doc in categories:
    print(doc)
```

The result of the previous code will look like this:

```
>>>
{'count': 181.0, 'sentiment': 4.0}
{'count': 177.0, 'sentiment': 0.0}
{'count': 139.0, 'sentiment': 2.0}
>>>
```

We can filter the result before the grouping with the cond attribute in Mongo shell or conition in PyMongo. This is analogous to the WHERE statement in SQL:

```
cond: { via: "kindle2" },
```

The group function in PyMongo will look as follows:

```
tweets.group(key={"sentiment":1},
        condition={"via": "kindle2" },
        initial={"count": 0},
        reduce="function(obj,    prev)
                {prev.count++;}")
```

Aggregation framework

The MongoDB aggregation framework is an easy way to get aggregated values and works fine with **sharding** without having to use **MapReduce** (see Chapter 13, *Working with MapReduce*). Aggregation framework is flexible, functional, and simple to implement operation pipelines and computational expressions. Aggregation Framework uses a declarative JSON format implemented in C++ instead of JavaScript, which improves the performance. The aggregate method prototype is shown here:

```
db.collection.aggregate( [<pipeline>] )
```

In the following code, we can see a simple counting by grouping the `sentiment` field with the `aggregate` method. In this case, the pipeline is only using the `$group` operator:

```
from pymongo import MongoClientcon = MongoClient()
db = con.Corpus
tweets = db.tweets
results = tweets.aggregate([
        {"$group": {"_id": "$sentiment", "count": {"$sum": 1}}}
    ])
for doc in results["result"]:
    print(doc)
```

In the next screenshot, we can see the result of the aggregation by grouping:

```
>>> ============================== RESTART ==============================
>>>
{'count': 139, '_id': 2}
{'count': 177, '_id': 0}
{'count': 181, '_id': 4}
>>>
```

We can find the documentation of the aggregation framework at `http://docs.mongodb.org/manual/reference/aggregation/`.

Pipelines

In a pipeline, we will process a stream of documents where the original input is a collection and the final output is a result document. The pipeline has a series of operators that filter or transform data, and it can generate a new document or a filtered out document.

The main pipeline operators are here:

- **$match**: This filters documents, uses existing query syntax and no geospatial operations or `$where`
- **$group**: This groups documents by an id and can use all the computational expressions like `$max`, `$min`, and so on
- **$unwind**: This operates on an array field, yields documents for each array value, and also complements `$match` and `$group`
- **$sort**: This sorts the documents by one or more fields
- **$skip**: This skips over the documents in the pipeline
- **$limit**: This restricts the number of documents in an aggregation pipeline

In the code here, we can see the aggregation with a pipeline using the $group, $sort, and $limit operators:

```
from pymongo import MongoClientcon = MongoClient()
db = con.Corpus
tweets = db.tweets
results = tweets.aggregate([
        {"$group": {"_id": "$via",
                    "count": {"$sum": 1}}},
        {"$sort": {"via":1}},
        {"$limit":10},
    ])
for doc in results["result"]:
    print(doc)
```

In the following screenshot, we can see the result of the aggregation using a pipeline with multiple operators:

```
>>> ============================ RESTART ============================
>>>
{'count': 1, '_id': 'fred wilson'}
{'count': 8, '_id': 'warren buffet'}
{'count': 1, '_id': 'aapl'}
{'count': 2, '_id': 'mashable'}
{'count': 1, '_id': 'hitler'}
{'count': 1, '_id': 'yankees'}
{'count': 1, '_id': 'republican'}
{'count': 7, '_id': 'exam'}
{'count': 1, '_id': 'world cup'}
{'count': 5, '_id': 'viral marketing'}
>>>
```

Expressions

Expressions produce output documents based on calculations performed on input documents. They are stateless and are only used in in the aggregation process.

The $group aggregation operations are here:

- **$max**: This returns the highest value in the group
- **$min**: This returns the lowest value in the group
- **$avg**: This returns the average of all the group values
- **$sum**: This returns the sum of all values in the group
- **$addToSet**: This returns an array of all the distinct values for a certain field in each document in that group

We can also find other kinds of operators depending on its data type:

- **Boolean**: $and, $or, and $not
- **Arithmetic**: $add, $divide, $mod, $multiply, and $substract
- **String**: $concat, $substr, $toUpper, $toLower, and $strcasecmp
- **Conditional**: $cond and $ifNull

In the code here, we will use the aggregate method with the $group operator; and in this case, we will use multiple operations such as $avg, $max, and $min:

```python
from pymongo import MongoClient
con = MongoClient()
db = con.Corpus
tweets = db.tweets
results = tweets.aggregate([
        {"$group": {"_id": "$via",
                    "avgId": {"$avg": "$id"} ,
                    "maxId": {"$max": "$id"} ,
                    "minId": {"$min": "$id"} ,
                    "count": {"$sum": 1}}}
    ])
for doc in results["result"]:
    print(doc)
```

Here we can see the result of $group using multiple operators:

```
>>> =============================== RESTART ===============================
>>>
{'count': 7, 'avgId': 1065.857142857143, '_id': 'exam', 'maxId': 2195, 'minId': 218}
{'count': 1, 'avgId': 226.0, '_id': 'republican', 'maxId': 226, 'minId': 226}
{'count': 1, 'avgId': 1025.0, '_id': 'world cup', 'maxId': 1025, 'minId': 1025}
{'count': 1, 'avgId': 2398.0, '_id': 'yankees', 'maxId': 2398, 'minId': 2398}
{'count': 1, 'avgId': 14045.0, '_id': 'aapl', 'maxId': 14045, 'minId': 14045}
{'count': 1, 'avgId': 2296.0, '_id': 'hitler', 'maxId': 2296, 'minId': 2296}
. . .
```

Aggregation framework has some limitations such as the document size limit is 16 MB, and there are some field types unsupported (**Binary**, **Code**, **MinKey**, and **MaxKey**).

About sharding support, MongoDB analyzes pipeline, and forwards operations up to $group or $sort to shards. Then, it combines shard server result and returns them. Due to this, it is recommended to use $match and $sort as early as possible into the pipeline.

Summary

In this chapter, we explored the basic operations and functions of MongoDB directly from the Mongo shell and from a graphic interface (Robomongo). We also performed a data preparation of a CSV dataset with OpenRefine, and turned it into a well-formatted JSON dataset. Finally, we presented an introduction to data processing with the aggregation framework, which is a faster alternative to MapReduce for common aggregations. We introduced the basic operators used in the pipelines and the expressions supported by the aggregation framework.

In the next chapter, we will explore the MapReduce functionality of MongoDB, and we will create a word cloud in D3 with the most frequent words in positive tweets.

13
Working with MapReduce

MongoDB is a document-based database used to tackle a large amount of data by companies such as **Forbes**, **Bit.ly**, **Foursquare**, **Craigslist**, and so on. In Chapter 12, *Data Processing and Aggregation with MongoDB*, we learned how to perform the basic operations and aggregations with MongoDB. In this chapter, we will learn how MongoDB implements a **MapReduce** programming model using **Jupyter** and **PyMongo**.

In this chapter, we will cover the following topics:

- An overview of MapReduce
- Programming model
- Using MapReduce with MongoDB
- Filtering the input collection
- Grouping and aggregation
- The most common words in Tweets

 In the following link, we can find a list of production deployments of MongoDB:
https://www.mongodb.com/industries

An overview of MapReduce

MapReduce is a programming model for large-scale distributed data processing, inspired by the map and reduce functions of functional programming languages such as **Lisp**, **Haskell**, and **Python**. One of the most important features of MapReduce is that it allows us to hide the low-level implementation, such as message passing or synchronization, from users and split a problem into many partitions. This is a great way to make the parallelization of data processing easy, without any need for communication between the partitions.

The original Google paper MapReduce: *Simplified Data Processing on Large Clusters*, can be found in the following link:
`http://research.google.com/archive/mapreduce.html`

MapReduce became mainstream because of **Apache Hadoop**, which is an open source framework that was derived from Google's MapReduce paper. MapReduce allows us to process massive amounts of data in a distributed cluster. In fact, there are many implementations of the MapReduce programming model; some of them are shown in the following list. It is important to say that MapReduce is not an algorithm, is just a part of a high performance infrastructure that provides a lightweight way to run a program on a lot of parallel machines.

Some of the most popular implementations of MapReduce are listed here:

- **Apache Hadoop**: Probably the most famous implementation of Google's MapReduce model, based on Java with an excellent community and vast ecosystem. Find more information in the following link:
  `http://hadoop.apache.org/`
- **MongoDB**: A document-oriented database that provides MapReduce operations. Find more information in the following link:
  `http://docs.mongodb.org/manual/core/map-reduce/`
- **MapReduce-MPI Library**: A MapReduce implementation that runs on top of the **MPI** (**Message Passing Interface**) standard. Find more information in the following link:
  `http://mapreduce.sandia.gov/`

Message passing is a technique used in concurrent programming to provide synchronization between processes, similar to a traffic light control system. MPI is a standard for message passing implementation. Find more information about MPI in the following link:
`http://en.wikipedia.org/wiki/Message_Passing_Interface`

Programming model

MapReduce provides us an easy way to create parallel programs without concern for message passing or synchronization. This can help us to perform complex aggregation tasks or searches. As we can observe in the following screenshot, MapReduce may work with less organized data (such as noisy text or schemaless documents) than the traditional relational databases. However, the programming model is more procedural, which means that the user must have some programming skills, such as in Java, Python, JavaScript, or C. MapReduce requires two functions: the Map function, which creates a list of key-value pairs; and Reduce, which iterates over each value and then applies some process (merge or summarization) to get an output.

The data could be split into several nodes (**sharding**); in this case, we will need a partition function. This partition function will be in charge of sorting and load balancing. In MongoDB, we can work over sharded collections automatically without any configuration:

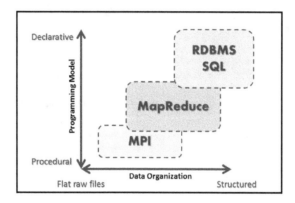

Using MapReduce with MongoDB

MongoDB provides us with a MapReduce command, and in the following diagram we can observe the life cycle of the MapReduce process in MongoDB. We start with a Collection or a Query; each document in the collection will call the map function. Then, with the emit function, we will create an intermediate hash-map (see the following diagram) with a list of pairs (key-value).

Next, the `reduce` function will iterate the intermediate hash-map and will apply some operations to all values of each key. Finally, the process will create a brand new collection with the output. The `map/reduce` functions in MongoDB will be programmed with JavaScript:

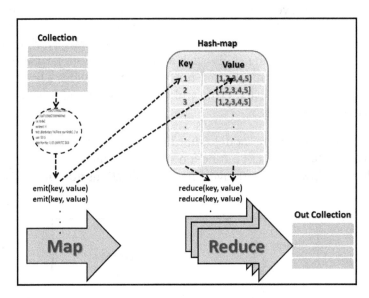

Find the reference documentation for MapReduce with MongoDB from the following link:
`http://docs.mongodb.org/manual/core/map-reduce/`

Map function

The `map` function will call the `emit` function one or more times (see the previous diagram). We can access all the attributes of each document in the collection with the `this` keyword. The intermediate hash-map contains only unique keys, so if the `emit` function sends a key that is already in the hash-map, the value is going to be inserted in a list of values. Each record in the hash-map will look like this: `key:One, value:[1,2,3,...]`.

Here, we can see a sample code of the `map` function:

```
function(){
  emit(this._id, {count: 1});
}
```

Reduce function

The `reduce` function will receive two arguments: the `key` and `values` (one or a list of values). This function is called for each record in the hash-map.

In the following code, we can see a sample `reduce` function. In this case, the function returns the total count for each key:

```
function(key, values) {
    total = 0;
    for (var i = 0; i < values.length; ++i) {
        total += values[i].count;
    };
    return {count: total};
}
```

See the *Using mongo shell* section to see how this function works.

Using mongo shell

Mongo shell provides a wrapper method for the `mapReduce` command. The `db.collection.mapReduce()` method must receive three parameters: the `map` function, the `reduce` function, and the name of the collection where the output is going to be stored, as shown in the following command:

```
db.collection.mapReduce(map, reduce, {out:"OutCollection"})
```

In this example, we will use the tweets collection that we created in Chapter 12, *Data Processing and Aggregation with MongoDB*, in the *Inserting Documents with PyMongo* section with the `id`, `via`, `sentiment`, `text`, `user`, and `date` attributes. This example will count how many times each unique element of the `via` attribute appears in the collection.

First, we need to define the `map` function in the `mapTest` variable:

```
mapTest = function(){
    emit(this.via, 1);
}
```

Then, we need to define the `reduce` function in the `reduceTest` variable:

```
reduceTest = function(key, values) {
        var res = 0;
            values.forEach(function(v){ res += 1})
        return {count: res};
            }
```

The mongo shell will look like this:

```
> mapTest = function(){ emit(this.via, 1); }
function (){ emit(this.via, 1); }
> reduceTest = function(key, values) {
...
... var res = 0;
... values.forEach(function(v){ res += 1})
...
... return {count: res};
... }
function (key, values) {

var res = 0;
values.forEach(function(v){ res += 1})

return {count: res};
}
>
```

Now, we need to define `Corpus` as the default database:

```
use Corpus
```

Next, we will use the `mapReduce` method, sending the `mapTest` and `reduceTest` functions and defining a new collection, results, to store the output:

```
db.tweets.mapReduce(mapTest, reduceTest, {out:"results"})
```

Finally, we will retrieve all the documents of the results collection with the `find` method:

```
db.results.find()
```

In the following screenshot, we can see the results of the `mapReduce` command in the mongo shell and the retrieved collection (results) with the aggregated data (count) of the `via` attribute:

```
> use Corpus
switched to db Corpus
> db.tweets.mapReduce(mapTest,reduceTest, {out:"results"})
{
        "result" : "results",
        "timeMillis" : 135,
        "counts" : {
                "input" : 497,
                "emit" : 497,
                "reduce" : 59,
                "output" : 80
        },
        "ok" : 1,
}
> db.results.find()
{ "_id" : 40, "value" : { "count" : 4 } }
{ "_id" : 50, "value" : { "count" : 6 } }
{ "_id" : "Bobby Flay", "value" : { "count" : 8 } }
{ "_id" : "Danny Gokey", "value" : { "count" : 4 } }
{ "_id" : "Malcolm Gladwell", "value" : { "count" : 11 } }
{ "_id" : "aapl", "value" : 1 }
{ "_id" : "aig", "value" : { "count" : 7 } }
{ "_id" : "at&t", "value" : { "count" : 15 } }
{ "_id" : "bailout", "value" : 1 }
{ "_id" : "baseball", "value" : { "count" : 6 } }
{ "_id" : "bing", "value" : 1 }
{ "_id" : "booz allen", "value" : { "count" : 3 } }
{ "_id" : "car warranty call", "value" : { "count" : 2 } }
{ "_id" : "cheney", "value" : { "count" : 5 } }
{ "_id" : "china", "value" : { "count" : 6 } }
{ "_id" : "comcast", "value" : { "count" : 4 } }
{ "_id" : "dentist", "value" : { "count" : 17 } }
{ "_id" : "driving", "value" : 1 }
{ "_id" : "east palo alto", "value" : { "count" : 4 } }
{ "_id" : "eating", "value" : { "count" : 12 } }
Type "it" for more
> _
```

 For a complete reference of the `mapReduce` command, use the following link:
http://bit.ly/13Yh5Kg

Using Jupyter

In this section, we will use Jupyter Notebook to execute a `mapReduce` command from a Notebook interface. First, we will open and connect Jupyter Notebook with the local MongoDB.

Now, we will execute Anaconda Launcher and select the **Jupyter Notebook** (see the following screenshot), click on **New/Notebook/Python 2**, and see a brand new Notebook in the browser:

Now, we will see the **Notebook** window, where we can insert the the code in the inline **In[1]** and then we will type the mapReduce code. Finally, we will click on the **Play** button, then we will see the result in the area below, as is shown in the next screenshot. To install pymongo or any other library in Jupyter, we may use the pip instruction, as shown here:

```
!pip install pymongo
```

For a complete reference for the Jupyter project, see this link: http://jupyter.org/

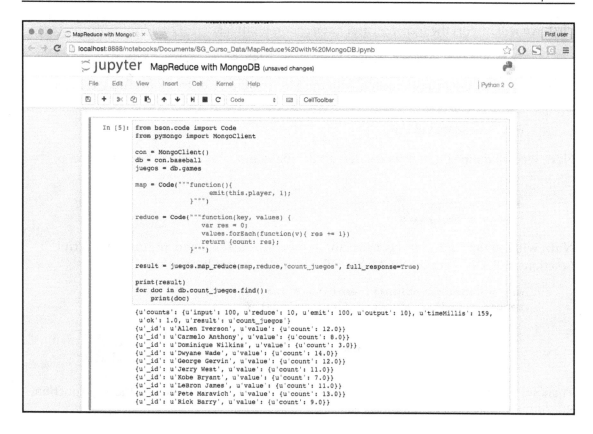

Using PyMongo

With mongo shell or Jupyter, we may run a MapReduce process easily. However, we will normally need to call the MapReduce process as a part of a bigger transaction. Then we need to implement a MapReduce wrapper in an external programming language. In this case, we will use `pymongo` to call the `mapReduce` command from Python.

In this example, we will use the tweets collection and will count how many times each via appears. See `Chapter 12`, *Data Processing and Aggregation with MongoDB*, in the *Inserting Documents with PyMongo* section for details about the creation of the `tweets` collection.

First, we will import the `pymongo` and `bson.code` modules:

```
from pymongo import MongoClientfrom bson.code import Code
```

Then, we will establish a connection with the MongoDB service to the default localhost and port 27017:

```
con = MongoClient()
```

Next, we will define `Corpus` as the default database and `tweets` as a shortcut object of `db.tweets`:

```
db = con.Corpus
tweets = db.tweets
```

Now, will use the `Code` object constructor to represent the `map` and `reduce` JavaScript functions in BSON, because the MongoDB API methods use JavaScript:

```
map = Code("function(){ emit(this.via, 1); }")
reduce = Code("""function(key, values) {
            var res = 0;
            values.forEach(function(v){ res += 1})
            return {count: res};
            }""")
```

Then, we will use the `map_reduce` function, providing three parameters, the `map` function, the `reduce` function, and a defined `via_count` as the output collection:

```
result = tweets.map_reduce(map, reduce, "via_count")
print(result)
```

Finally, we retrieve all the documents in the `via_count` collection with the `find` function:

```
for doc in db.via_count.find():
    print(doc)
```

In the following screenshot, we can observe the results of the preceding code:

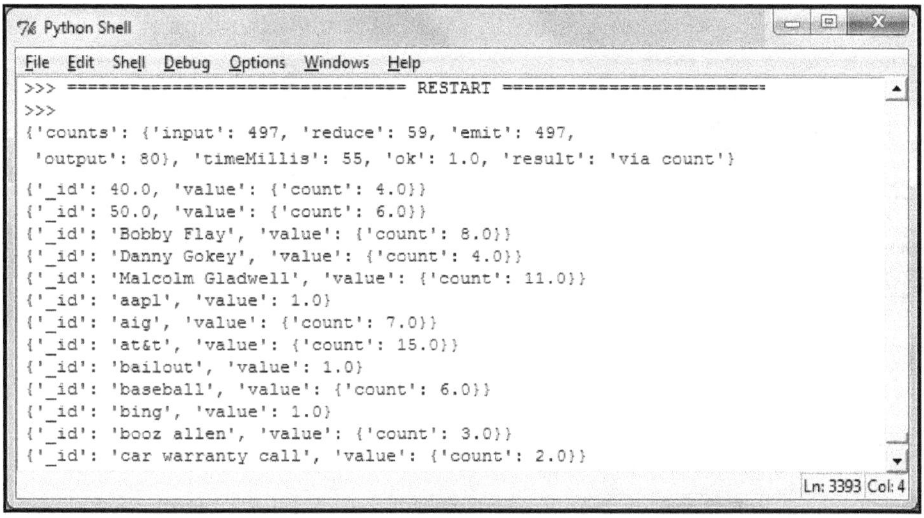

Filtering the input collection

Sometimes, we don't need the entire collection for our MapReduce process. Because of this, the mapReduce command provides us with optional parameters to filter the input collection.

The query parameter allows us to apply criteria using the query operators to filter the documents input into the map function. In the following code, we will filter the documents in the collection and only include the documents where the number attribute is greater than 10 ("$gt":10):

```
collection.map_reduce(map_function,
                      reduce_function,
                      "output_collection",
                      query={"number":{"$gt":10}})
```

The **Query Operators** used in the MapReduce query parameter are the same query selectors seen in `Chapter 12`, *Data Processing and Aggregation with MongoDB*, which perform simple queries. In the following table, we present the most common operators and their equivalents in SQL:

Mongo operators	SQL operators
$gt	>
$gte	>=
$in	IN
$lt	<
$lte	<=
$and	AND
$or	OR

`limit` is an optional parameter of the `map_Reduce` command, which helps us to define the maximum number of documents retrieved by the query. In the following code, we define the `limit` of documents retrieved as a maximum of `10`:

```
collection.map_reduce(map_function,
                      reduce_function,
                      "output_collection",
                      limit = 10)
```

Find a complete list of MongoDB operators in the following link:
http://docs.mongodb.org/manual/reference/operator/

Grouping and aggregation

In the following example, we will perform **grouping** and **aggregation** in order to get statistics (**Sum**, **Max**, **Min**, and **Average**) about NBA players and which number of points they scored. First, the `map` function will send the name of the player and the number of points scored for each game. The `map` function will look like this:

```
function(){emit(this.player, this.points); }
```

Then, we can perform all the aggregation functions simultaneously using the `sum` method from the JavaScript `Array` object and the `max/min` functions of the JavaScript `Math` object. The `reduce` function will look like this:

```
function(key, values) {
    var explain = {total:Array.sum(values),
                    max:Math.max.apply(Math, values),
                    min:Math.min.apply(Math, values),
                    avg:Array.sum(values)/values.length}
    return explain
}
```

For this example, we will create synthetic data, randomly mixing the names of the 10 players and assigning a random score between 0 and 100. Then we will insert the data into a MongoDB collection called `Games`. The complete code is shown here:

```
import random as ran
import pymongo
con = pymongo.Connection()
db = con.basketball
games = db.games
players = ["LeBron James",
            "Allen Iverson",
            "Kobe Bryant",
            "Rick Barry",
            "Dominique Wilkins",
            "George Gervin",
            "Dwyane Wade",
            "Jerry West",
            "Pete Maravich",
            "Carmelo Anthony"]
for x in range(100):
    games.insert({ "player" : players[ran.randint(0,9)],
                    "points" : ran.randint(0,100)})
```

The `Games` collection will look like this. We can observe that a `player` can appear several times with different point scores:

```
{ "_id" : { "$oid" : "5206caef9dd27c1964b1d648"} , "player" : "LeBron James" , "points" : 40}
{ "_id" : { "$oid" : "5206caef9dd27c1964b1d649"} , "player" : "Rick Barry" , "points" : 6}
{ "_id" : { "$oid" : "5206caef9dd27c1964b1d64a"} , "player" : "George Gervin" , "points" : 0}
{ "_id" : { "$oid" : "5206caef9dd27c1964b1d64b"} , "player" : "Kobe Bryant" , "points" : 56}
{ "_id" : { "$oid" : "5206caef9dd27c1964b1d64c"} , "player" : "Pete Maravich" , "points" : 4}
{ "_id" : { "$oid" : "5206caef9dd27c1964b1d64d"} , "player" : "Dwyane Wade" , "points" : 65}
{ "_id" : { "$oid" : "5206caef9dd27c1964b1d64e"} , "player" : "Pete Maravich" , "points" : 55}
{ "_id" : { "$oid" : "5206caef9dd27c1964b1d64f"} , "player" : "Dwyane Wade" , "points" : 45}
{ "_id" : { "$oid" : "5206caef9dd27c1964b1d650"} , "player" : "Allen Iverson" , "points" : 66}
{ "_id" : { "$oid" : "5206caef9dd27c1964b1d651"} , "player" : "Rick Barry" , "points" : 18}

    . . .
```

Finally, we will perform the MapReduce process using the `map/reduce` functions seen in the beginning of this section, implemented in `pymongo`. We will store the output in the `_result` collection. Here, we can see the complete code:

```python
from pymongo import MongoClientfrom bson.code import Code
con = MongoClient()db = con.basketball
games = db.games
map = Code("""function(){
                emit(this.player, this.points);
            }""")
reduce = Code("""function(key, values) {
                var explain = {total:Array.sum(values),
                        max:Math.max.apply(Math, values),
                        min:Math.min.apply(Math, values),
                        avg:Array.sum(values)/values.length}
                return explain;
            }""")
result = games.map_reduce(map, reduce, "_result")
print(result)
```

The result of the grouping and aggregation will look like this:

```
>>> ============================ RESTART ================================
>>>
Collection(Database(Connection('localhost', 27017), 'baseball'), '_result')
{'_id': 'Allen Iverson', 'value': {'max': 66.0, 'total': 310.0, 'avg': 34.44444444444444, 'min': 9.0}}
{'_id': 'Carmelo Anthony', 'value': {'max': 91.0, 'total': 473.0, 'avg': 47.3, 'min': 1.0}}
{'_id': 'Dominique Wilkins', 'value': {'max': 98.0, 'total': 545.0, 'avg': 60.55555555555556, 'min': 20.0}}
{'_id': 'Dwyane Wade', 'value': {'max': 95.0, 'total': 834.0, 'avg': 55.6, 'min': 15.0}}
{'_id': 'George Gervin', 'value': {'max': 81.0, 'total': 235.0, 'avg': 47.0, 'min': 0.0}}
{'_id': 'Jerry West', 'value': {'max': 98.0, 'total': 645.0, 'avg': 58.63636363636363, 'min': 9.0}}
{'_id': 'Kobe Bryant', 'value': {'max': 95.0, 'total': 497.0, 'avg': 45.18181818181818, 'min': 0.0}}
{'_id': 'LeBron James', 'value': {'max': 100.0, 'total': 546.0, 'avg': 49.63636363636363, 'min': 3.0}}
{'_id': 'Pete Maravich', 'value': {'max': 97.0, 'total': 562.0, 'avg': 43.23076923076923, 'min': 4.0}}
{'_id': 'Rick Barry', 'value': {'max': 98.0, 'total': 781.0, 'avg': 48.8125, 'min': 6.0}}
>>> |
```

All the code and datasets for this chapter may be found in the author's GitHub repository, in the following link:
`https://github.com/hmcuesta/PDA_Book/Chapter13`

Counting the most common words in tweets

In this example, we will develop a simple application that counts the number of occurrences of each word in positive tweets. First, we will split each tweet into words. Then we remove all the URLs (`http://...`) and twitter users (`@...`). Next, we will remove all the words with three or less characters (like *the, why, she, him*, and so on). Finally, the counting word frequencies. In the following code, we can see the JavaScript `map` function spliting words from tweets:

```
function(){
    this.text.split(' ').forEach(
        function(word){
            var txt = word.toLowerCase();
            if(!(/^@/).test(txt) &&
                 txt.length >= 3 &&
               !(/^http/).test(txt)){
                  emit(txt,1)
            }
        }
    }
}
```

The input will look like this:

```
'text': '@SomeUsr After using LaTeX a lot any other typeset mathematics
just looks greate. http://www.latex.org',
```

The output will look like this. The `emit` function will be called for each word:

```
["after", "using", "latex",  "other", "typeset", "mathematics", "
just", "looks", "great"]
```

In the following code, we can see the JavaScript `reduce` function to get the frequency of occurrence of each word:

```
function(key, values) {
            var res = 0;
            values.forEach(function(v){ res += 1})
            return {count: res};
}
```

In Chapter 11, *Working with Twitter Data*, we talked about how a bag-of-words model is a common method for document classification, using the frequency of occurrences of each word as a feature for the classifier.

For this example, we will use the Corpus database and the tweets collection created in Chapter 12, *Data Processing and Aggregation with MongoDB*, in the *Inserting Documents with PyMongo* section. Each document in the tweets collection will look like this:

```
{'via': 'latex',
 'sentiment': 4,
 'text': '@SomeUsr After using LaTeX a lot any other typeset
mathematics just looks greate. http://www.latex.org',
 'user': 'yomcat',
 'date': 'Sun Jun 14 04:31:28 UTC 2009',
 '_id': ObjectId('51ed71359dd27c0b94666696'),
 'id': 1407        1}
```

In the following code, we will perform a map_reduce method, querying only the positive tweets (sentiment = 4) as an input collection:

```python
from pymongo import MongoClientfrom bson.code import Code
import csv
con = MongoClient()db = con.Corpus
tweets = db.tweets
map = Code("""function(){
                this.text.split(' ').forEach(
                    function(word){
                        var txt = word.toLowerCase();
                        if(!(/^@/).test(txt) &&
                            txt.length > 3 &&
                             !(/^http/).test(txt)){
                                emit(txt,1)
                        }
                    }
                )
            }""")
reduce = Code("""function(key, values) {
                var res = 0;
                 values.forEach(function(v){ res += 1})
                 return {count: res};
                 }""")
result = tweets.map_reduce(map,reduce, "TweetWords",
query={"sentiment":4})
```

The output collection will be stored in `TweetWords`; we may check the number of resulting words (2173) with this command:

```
db.runCommand( { count: TweetWords })
```

In the following image, we may see the `count` and content of the `TweetWords` collection:

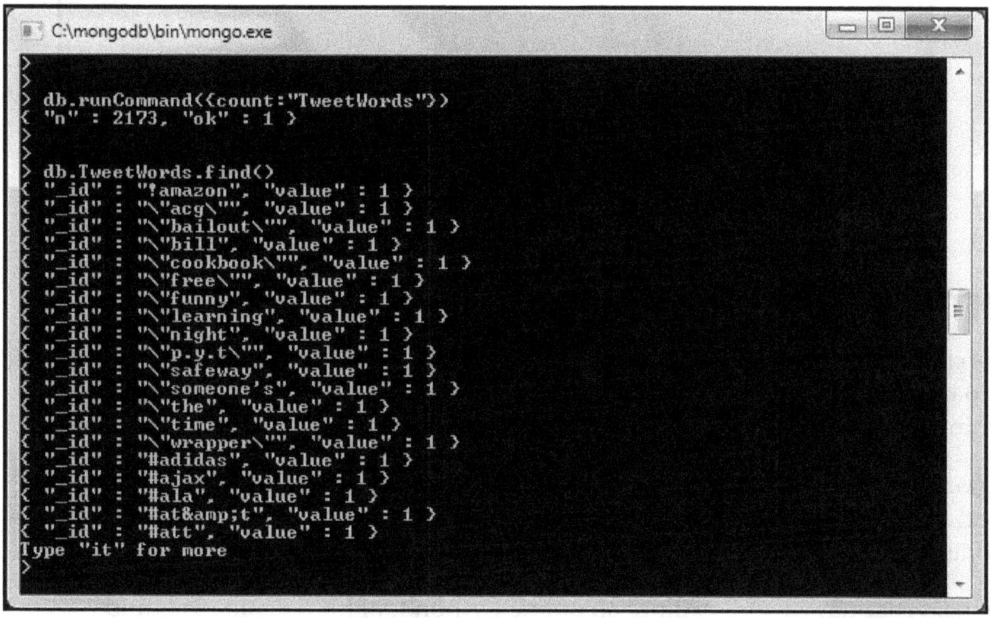

Now, for our visualization we need a csv file with the 50 most frequent words. In the following code, we perform a query of the `TweetWords` collection, sorting the results in descending order and limiting the output to only the first 50 documents. Finally, we will store the output in the `data.csv` file:

```
with open("data.csv", "w") as f:
    f_csv = csv.writer(f, delimiter=',')
    f_csv.writerow(["text","size"])
    for doc in db.TweetWords.find()
                        .sort("value", direction = -1)
                        .limit(50):
        f_csv.writerow([doc["_id"],doc["value"]["count"]+30])
        print(doc)
```

We can see the output of the query in the following screenshot:

```
76 Python Shell
File  Edit  Shell  Debug  Options  Windows  Help
>>>
Collection(Database(Connection('localhost', 27017), 'Corpus'), 'TweetWords')
{'_id': 'love', 'value': {'count': 28.0}}
{'_id': 'good', 'value': {'count': 18.0}}
{'_id': 'just', 'value': {'count': 18.0}}
{'_id': 'with', 'value': {'count': 18.0}}
{'_id': 'have', 'value': {'count': 17.0}}
{'_id': 'night', 'value': {'count': 15.0}}
{'_id': 'from', 'value': {'count': 13.0}}
{'_id': 'nike', 'value': {'count': 13.0}}
                                                              Ln: 2882 Col: 4
```

Summary

In this chapter, we explored the basic concepts of the MapReduce programming model and how to implement common activities such as grouping, aggregation, counting, and summing in MongoDB.

MapReduce is a powerful tool for log analysis and data processing. In this chapter, we learned how to implement easy but powerful aggregation capabilities into Python using PyMongo and Jupyter.

In the next chapter, we will explore an online Python tool for data analysis and development called Wakari, and a data analysis library named Pandas.

14
Online Data Analysis with Jupyter and Wakari

In this chapter, we will introduce an online tool for data analysis called **Wakari**, in which we will set up a complete Python environment in just a few seconds. Then we will present some of the capabilities of Wakari through **Jupyter Notebook** by using the PIL and `pandas` libraries.

In this chapter, we will cover:

- Getting started with Wakari
- Getting started with Jupyter Notebook
- Introduction to image processing with PIL
- Introduction to data analysis with pandas
- Sharing your Notebook

Getting started with Wakari

Wakari is a cloud service for collaborative Python data analysis environments, created by **Continuum Analytics**. Wakari provides a powerful set of pre-configured Python environments built on top of **Anaconda**, which is a free Python distribution for large-scale data processing and scientific computing. Wakari uses a Jupyter GUI, which is a Python shell improved to write, debug, and test Python code for scientific computing.

Jupyter is an open source, interactive data science tool with support for 40 programming languages. Jupyter provides a terminal-based interface and a Wolfram-Matematica such as HTML notebook. In Wakari, we can use either the terminal console or Jupyter notebook.

Wakari helps us to set up a complete scientific Python environment without any local installation. This can be very convenient for learning purposes, because we may start coding right away, and the Anaconda distribution includes several of the most used libraries, such as `NumPy`, `SciPy`, `Matplotlib`, `PIL`, `pandas`, `Numba`, and so on.

In Wakari, we may use different kinds of terminals, such as **Python**, **Shell**, **IPython**, or **SSH**. However, in this chapter we will focus on the use of the **IPython notebook**.

The Jupyter Notebook is a rich web interface for coding. The notebook is a great tool for teaching and presenting Python code in an interactive interface. In this chapter, we will use the Jupyter Notebook included in Wakari and we will test some of its capabilities by implementing examples in **PIL** (**Python Image Library**) and Pandas.

> For more information about Jupyter, follow this link:
> `http://jupyter.org/`

Creating an account in Wakari

To start working with Wakari, we need to create an account or log in if we already have an account. We can create a new account in the following link:

`https://www.wakari.io/`

In the following screenshot, we can see the web form to register a new free account. In this chapter, we will work with the free account, which has some restrictions. However, we can find plans from 10 dollars, which can give us access via SSH and the capability to execute long-running jobs:

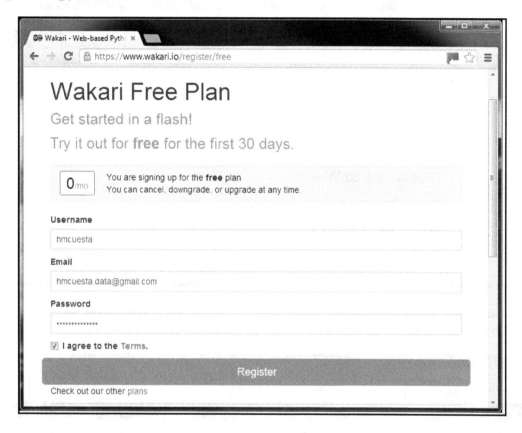

Once we log in to Wakari, the interface will look like the following screenshot, with the tabs on the right-hand side of the window for the terminals, the Jupyter Notebooks, and a **New Notebook** button.

On the left-hand side of the window, we may see the account **Path** with the resources (files and folders uploaded by the user):

If we click on the **Terminals** tab, we can add a new Python shell, Linux shell, or IPython shell. In the following screenshot, we can see a new **Python** shell:

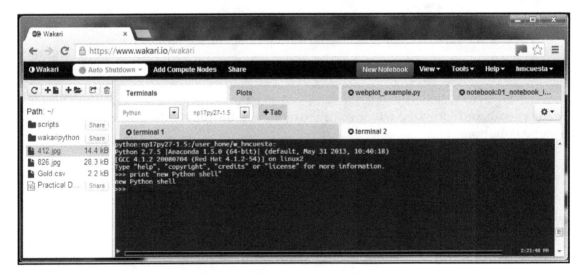

If we click on **Tools** and then select **Anaconda Environment Browser**, we can see a complete list of the installed packages and modules, as shown in the following screenshot:

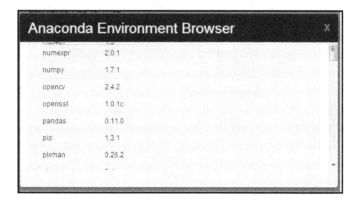

Getting started with IPython notebook

IPython evolved into the Jupyter project due to the proliferation of language-agnostic components, so the Anaconda installation has been moved to use Jupyter instead of IPython. However, Wakari still implements IPython for the Notebooks and shell. The IPython notebook (NB) is a web interface for our Python code. NB is based on a JSON format shareable and portable in .pynb file format.

To start with a blank notebook, we will click on the **New Notebook** button. In the following screenshot, we can see how to change the name by clicking on the **Untitled0** label; then we will rename the notebook:

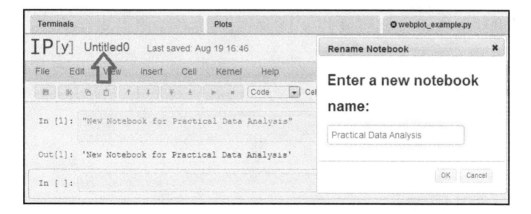

The NB will have access to all resources (text files, screenshots, and so on) in the path. We can upload text files, screenshots, and other content to the Wakari platform by clicking on the **Upload** icon (see the arrow in the following screenshot), then we will select the files, and finally, we will click on the **Upload Files** button, as shown in the following screenshot:

Finally, to run the code of our NB, we will click on the play icon (see the arrow in the following screenshot). We will get a numbered output for each of our input codes, as shown in the following screenshot. We may code several lines in the same input (In [1]), which we call cells, and as a result we can see the plot in the output (Out [1]). We also have access to all the modules included in the Anaconda distribution and all the resources in the path:

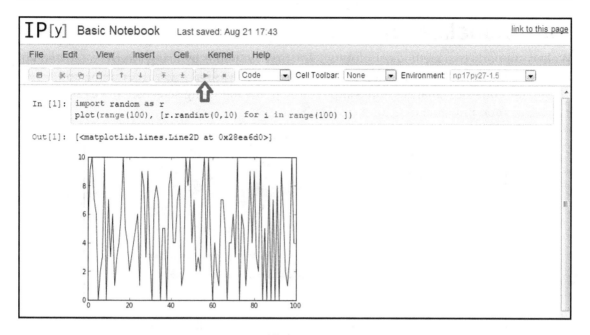

When we need to save the progress of our NB, we will click on the **File** menu and then select **Save**. If we need to create a local copy of our NB, we can click on the **File** menu and then click on **Download as**. Next, we can choose either the NB file (.ipynb) or the raw Python code (.py).

We can find more information about the IPython Notebook and its installation from its website:
`http://jupyter.readthedocs.io/en/latest/install.html`

Data visualization

Wakari supports two methods of plotting. The first method is by using `matplotlib` and all its capabilities. `PyLab` is just a wrapper for modules such as `matplotlib`, `numpy`, and `scipy`, for numerical analysis and computation. In the following screenshot, we can see a `plot_surface` implementing an **Axes3D** object:

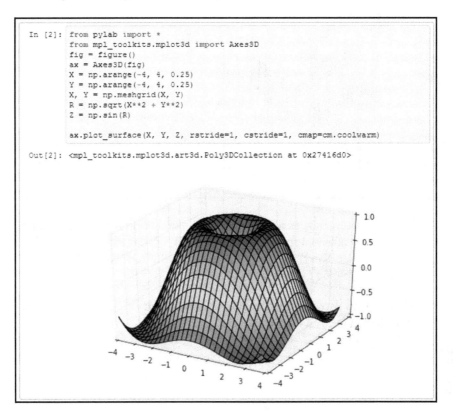

We can find more information about `matplotlib` from its website: http://matplotlib.org/

The second method for plotting in Wakari is through their custom plotting library, `webplot` (still in development), which create SVG graphics, currently just supporting line plots and scatter plots. In the following screenshot, we observe an example of a scatterplot of random points using `webplot`:

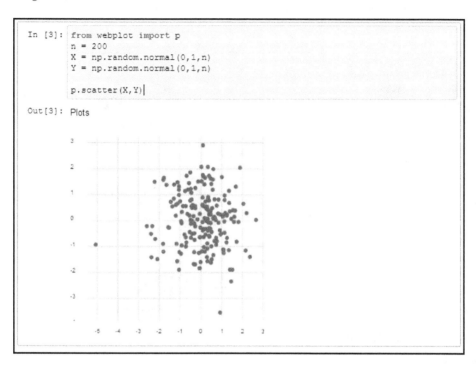

Introduction to image processing with PIL

The goal of this chapter is to present some of the pre-installed capabilities of Wakari. In this section, we will explore some of the basic functions of **PIL** (**Python Image Library**), such as histograms, filters, operations, and transformations. We already installed and used PIL in Chapter 5, *Similarity-Based Image Retrieval*.

First, we will upload the images `412.jpg` (Dinosaur) and `826.jpg` (Land) to the path (see the arrow in the following screenshot). The images came from the Caltech-256 images-dataset used in Chapter 5, *Similarity-Based Image Retrieval*.

Opening an image

The first thing we need to start working is to import the `PIL` and `pylab` modules. Next, we will use the `open()` method of the `Image` object. Finally, we will visualize the image with the `imshow()` method of `pylab`. In the following screenshot, we can see the output of the code:

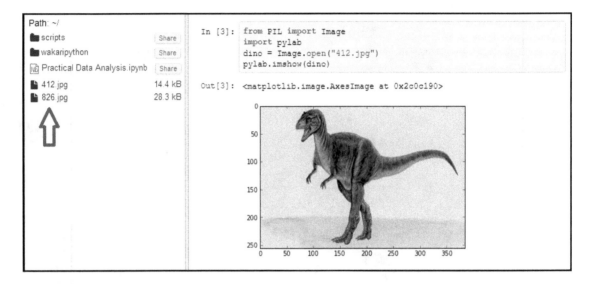

We can find more information about `PIL` from its website:
`http://www.pythonware.com/products/pil/`

Working with an image histogram

A **histogram** is a distribution of the frequency of the intensity of each pixel. `PIL` provides us with a histogram method, which will get the frequency of each tone of color. As our images are in **RGB (Red, Green, Blue)**, we will get an array of 768 values (256 tones * 3 colors).

Often, we will need the histogram of a grayscale image as it will be easier to work with only 256 values of gray intensity, instead of the full RGB color model. In `PIL`, we just add the parameter `"L"` to the `histogram()` method and the image will be treated as a grayscale image:

```
hist = land.histogram("L")
```

In the following screenshot, we will get the RGB histogram of the image (826.jpg) and we will plot the histogram() method using the hist() method of pylab:

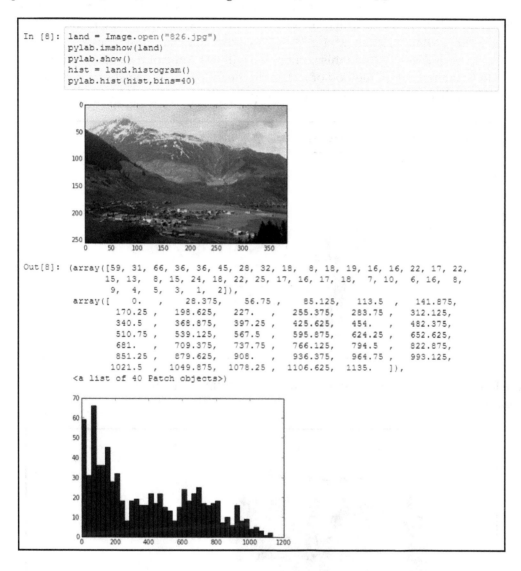

```
In [8]:  land = Image.open("826.jpg")
         pylab.imshow(land)
         pylab.show()
         hist = land.histogram()
         pylab.hist(hist,bins=40)
```

```
Out[8]:  (array([59, 31, 66, 36, 36, 45, 28, 32, 18,  8, 18, 19, 16, 16, 22, 17, 22,
                 15, 13,  8, 15, 24, 18, 22, 25, 17, 16, 17, 18,  7, 10,  6, 16,  8,
                  9,  4,  5,  3,  1,  2]),
          array([   0.   ,     28.375,     56.75 ,      85.125,    113.5  ,    141.875,
                   170.25 ,    198.625,    227.   ,     255.375,    283.75 ,    312.125,
                   340.5  ,    368.875,    397.25 ,     425.625,    454.   ,    482.375,
                   510.75 ,    539.125,    567.5  ,     595.875,    624.25 ,    652.625,
                   681.   ,    709.375,    737.75 ,     766.125,    794.5  ,    822.875,
                   851.25 ,    879.625,    908.   ,     936.375,    964.75 ,    993.125,
                  1021.5  ,   1049.875,   1078.25 ,    1106.625,   1135.   ]),
          <a list of 40 Patch objects>)
```

Filtering

The `filter()` method will return a copy of the image filtered by the given filter. We will use the `ImageFilter` object, which currently supports the following filters: BLUR, CONTOUR, DETAIL, EDGE_ENHANCE, EDGE_ENHANCE_MORE, EMBOSS, FIND_EDGES, SMOOTH, SMOOTH_MORE, and SHARPEN. In this section, we will test some of the common filters and plot them with the `imshow` method of `pylab`. In the following screenshot, we observe the BLUR filter applied to the image of the dinosaur:

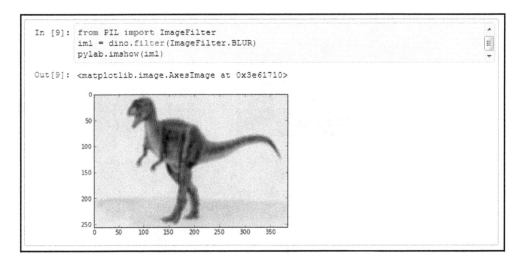

In the following screenshot, we observe the FIND_EDGES filter applied to the image of the dinosaur:

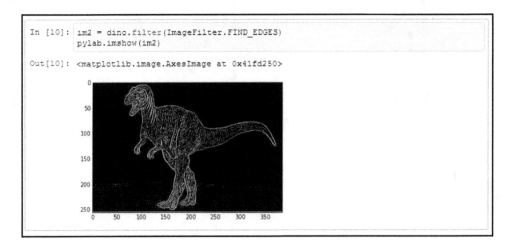

In the following screenshot, we observe the EDGES_ENHANCE_MORE filter applied to the image of the land:

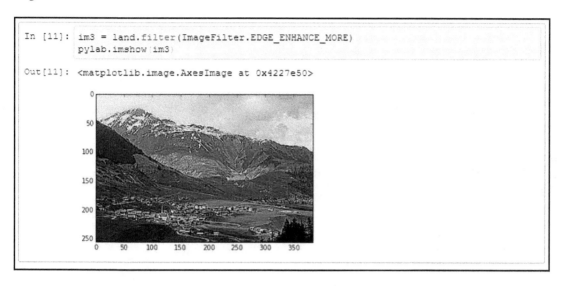

In the following screenshot, we observe the COUNTOUR filter applied to the image of the land:

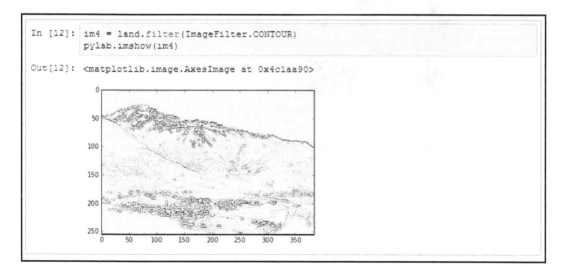

For reference documentation on the `ImageFilter` object, follow this link: `http://bit.ly/1fenKFq`

Operations

`PIL` includes some of the most common image processing operations ready to use with the `ImageOps` object.

In the following screenshot, we can see the dinosaur image using operation invert, which inverts each pixel value (photographic negative). We will use the `invert()` method from the `ImageOps` object included in the `PIL` library:

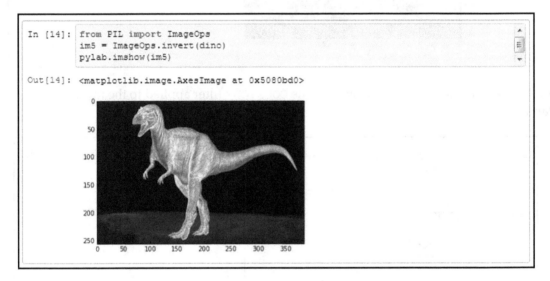

In the following screenshot, we can see the dinosaur image converted to grayscale:

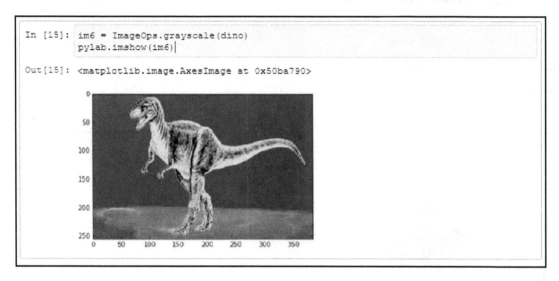

In the following screenshot, we can see the dinosaur image using a `solarize()` method, which inverts all pixel values above a given threshold:

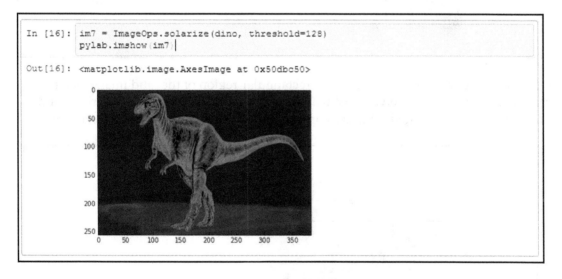

For reference documentation about the `ImageOps` object, follow this link:
`http://bit.ly/1741meW`

Transformations

PIL provides us with several methods for image transformations, such as `transform`, `transpose`, `crop`, and so on.

In the following screenshot, we see a rotated copy of the land image using the `transpose` method and we may use any of the following options: FLIP_LEFT_RIGHT, FLIP_TOP_BOTTOM, ROTATE_90, ROTATE_180, or ROTATE_270.

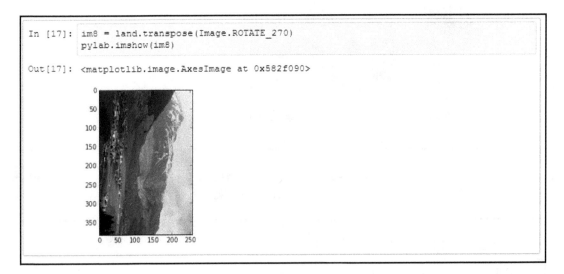

In the following screenshot, we can see a rectangular region of the land image using the `crop()` method, which receives a list with the pixel coordinates (left, upper, right, and lower). The `crop()` method returns a copy of the rectangular region from the image:

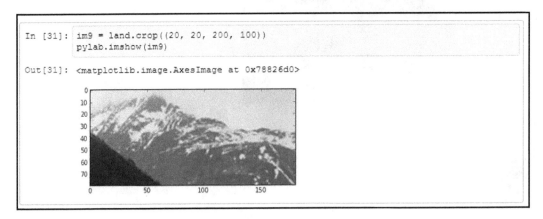

Getting started with pandas

The pandas library is a great library for data manipulation and analysis, written by Wes McKinney. The pandas library provides us with the optimized data structures, series and DataFrame, which are well-suited for descriptive statistics, indexing, and aggregation. Pandas is already installed in the Anaconda distribution used in Wakari. In this section, we will present the basic operations with pandas for Time Series and Multivariate data. We can find more information about pandas from its website:

```
http://pandas.pydata.org/
```

Working with Time Series

Time Series help us to understand the changes in a variable over time. Pandas include specific functionality in order to work with Time Series transparently. For this section, we need to upload the Gold.csv file used in Chapter 7, *Predicting Gold Prices*. The first five rows in the file will look like this:

```
date,price
1/31/2003,367.5
2/28/2003,347.5
3/31/2003,334.9
4/30/2003,336.8
5/30/2003,361.4
    .    .    .
```

We will load the Gold.csv file with the read_csv method (previously uploaded to the path of your account) and we will parse the dates just by activating the parse_date parameter (parse_dates=True). In the following screenshot, we can see that the result of loading is a DataFrame with a DatetimeIndex and a data column with the prices:

```
In [1]: import pandas as pd
        ts = pd.read_csv('Gold.csv', index_col=0, parse_dates=True)
        ts

Out[1]: <class 'pandas.core.frame.DataFrame'>
        DatetimeIndex: 125 entries, 2003-01-31 00:00:00 to 2013-05-31 00:00:00
        Data columns (total 1 columns):
        price    125  non-null values
        dtypes: float64(1)
```

Next, we will plot the Time Series just by calling the `plot()` method of our `DataFrame`. In the following image, we can see the gold prices from 2003 to 2013. The `plot()` method of the `DataFrame` is a wrapper of the `plt.plot()` method of `matplotlib`:

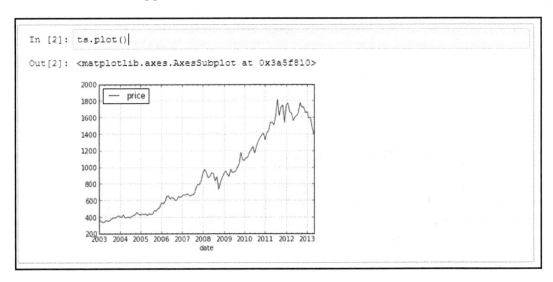

We can slice the time series just by specifying a range; in the case of the following screenshot, we just plot the records between 2006 and 2007 (`["2006":"2007"]`):

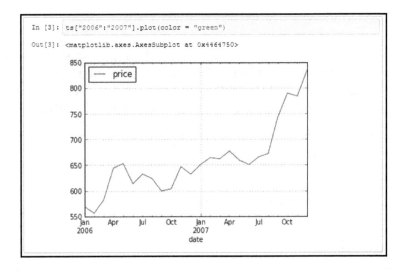

We may also define a specific date, `ts["2003/05/30"]`, or a month, `ts["2003/05"]`. Time Series can be also sliced between two dates using the `truncate` method:

```
ts.truncate(after = "05/30/2003")
```

The `pandas` library provides us with flexible resampling operations to perform frequency (Monthly, Yearly, Weekly, Daily, and so on) conversions. In the following screenshot, we will convert monthly data into annual data with the resample method. We will see a much smoother series in the `plot`:

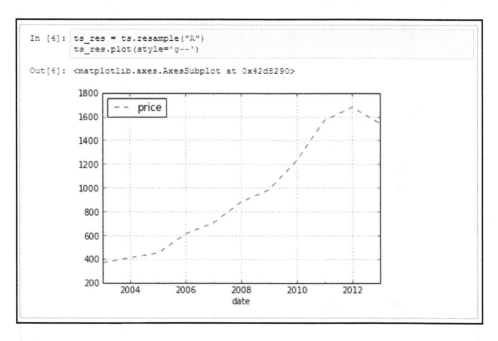

The `how` parameter of the `resample` method could be a custom function name or `numpy` array function that takes an array and produces aggregated data. For example, if we want only the max values, we will set the parameter as follows:

```
ts.resample("A", how=[np.max])
```

In the following screenshot, we will get three series: mean, max, and min. We will plot them in two different ways: the first is with the subplots=True option, which will display three different figures, and the second is a direct plot, in which we will see three lines in the same figure:

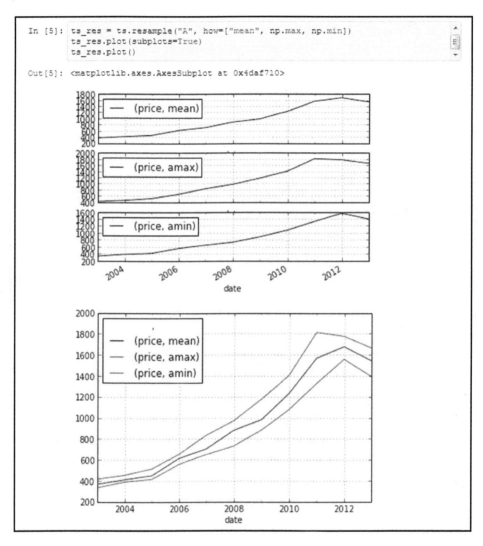

In the following link, we may find `pandas` Time Series documentation:
`http://pandas.pydata.org/pandas-docs/dev/timeseries.html`

Working with multivariate datasets with DataFrame

In this section, we will perform some descriptive statistics with a multivariate dataset using a `pandasDataFrame` object. In this section, we will use the `iris.csv` dataset; therefore, we need to upload the file into the Wakari path before we start working in our IPython Notebook. The iris flower dataset is probably the most used dataset for classification, with three categories (setosa, versicolour, virginica), four attributes (sepal length, sepal width, petal length, petal width), and 150 rows. We can download the iris dataset from the **UC Irvine Machine Learning Repository** at the following link:

`http://archive.ics.uci.edu/ml/datasets/Iris`

The first five records in the `iris.csv` file will look like this:

```
name,SepalLength,SepalWidth,PetalLength,PetalWidth
setosa,5.1,3.5,1.4,0.2
setosa,4.9,3,1.4,0.2
setosa,4.7,3.2,1.3,0.2
setosa,4.6,3.1,1.5,0.2
setosa,5,3.6,1.4,0.2
  .    .    .
```

First, we need to load the `iris.csv` file into a DataFrame using the `read_csv` method. Then we will plot the dataset using **RadViz**, which is a radial visualization that can help us to visualize multivariate data. The visualized attributes are presented as anchor points equally split around the perimeter of a circle, and in the following screenshot, we may see the **SepalLength**, **SepalWidth**, **PetalLength**, and **PetalWidth** anchors. The dataset instances (rows) are shown as points inside the circle and this visualization can be used as a classification technique.

In the following screenshot, we can see the plot of the iris dataset using the `radviz` method:

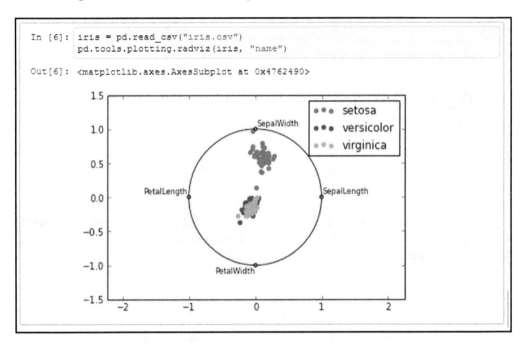

```
In [6]:  iris = pd.read_csv("iris.csv")
         pd.tools.plotting.radviz(iris, "name")

Out[6]:  <matplotlib.axes.AxesSubplot at 0x4762490>
```

The `pandas` library provides us with the head method (see following screenshot), which will get the first five records of our DataFrame, and the `tail` method, which will get the last five records:

```
In [7]:  iris.head()

Out[7]:
```

	name	SepalLength	SepalWidth	PetalLength	PetalWidth
0	setosa	5.1	3.5	1.4	0.2
1	setosa	4.9	3.0	1.4	0.2
2	setosa	4.7	3.2	1.3	0.2
3	setosa	4.6	3.1	1.5	0.2
4	setosa	5.0	3.6	1.4	0.2

We can get basic statistics from the `DataFrame` object with the `max`, `min`, and `mean` methods individually. But we can also get a summary of the DataFrame with the `describe` method, as shown in the following screenshot:

```
In [10]: iris.describe()
Out[10]:
```

	SepalLength	SepalWidth	PetalLength	PetalWidth
count	150.000000	150.000000	150.000000	150.000000
mean	5.843333	3.057333	3.758000	1.199333
std	0.828066	0.435866	1.765298	0.762238
min	4.300000	2.000000	1.000000	0.100000
25%	5.100000	2.800000	1.600000	0.300000
50%	5.800000	3.000000	4.350000	1.300000
75%	6.400000	3.300000	5.100000	1.800000
max	7.900000	4.400000	6.900000	2.500000

With a scatterplot, we see the correlation between two variables. However, when we have a multivariate dataset, the number of scatter plots increases. In these cases, we can use a scatterplot matrix in order to make it easier to plot the correlations of a dataset.

The `pandas` library provides us with the `scatter_matrix` method in `pandas.tools.plotting`, as shown in the following screenshot:

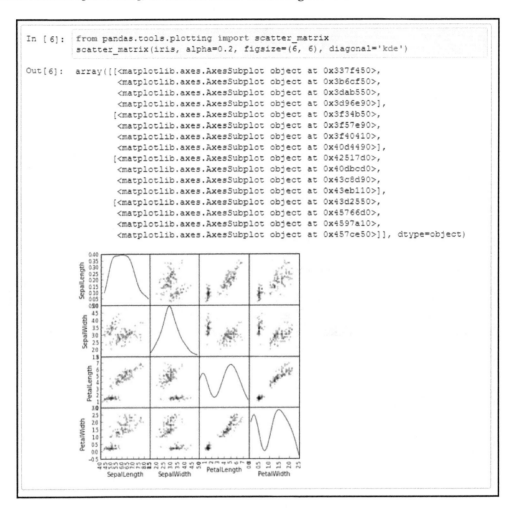

```
In [ 6]:  from pandas.tools.plotting import scatter_matrix
          scatter_matrix(iris, alpha=0.2, figsize=(6, 6), diagonal='kde')

Out[6]:   array([[<matplotlib.axes.AxesSubplot object at 0x337f450>,
                  <matplotlib.axes.AxesSubplot object at 0x3b6cf50>,
                  <matplotlib.axes.AxesSubplot object at 0x3dab550>,
                  <matplotlib.axes.AxesSubplot object at 0x3d96e90>],
                 [<matplotlib.axes.AxesSubplot object at 0x3f34b50>,
                  <matplotlib.axes.AxesSubplot object at 0x3f57e90>,
                  <matplotlib.axes.AxesSubplot object at 0x3f40410>,
                  <matplotlib.axes.AxesSubplot object at 0x40d4490>],
                 [<matplotlib.axes.AxesSubplot object at 0x42517d0>,
                  <matplotlib.axes.AxesSubplot object at 0x40dbcd0>,
                  <matplotlib.axes.AxesSubplot object at 0x43c8d90>,
                  <matplotlib.axes.AxesSubplot object at 0x43eb110>],
                 [<matplotlib.axes.AxesSubplot object at 0x43d2550>,
                  <matplotlib.axes.AxesSubplot object at 0x45766d0>,
                  <matplotlib.axes.AxesSubplot object at 0x4597a10>,
                  <matplotlib.axes.AxesSubplot object at 0x457ce50>]], dtype=object)
```

In the following link, we can find the Pandas DataFrame documentation:
`http://pandas.pydata.org/pandas-docs/dev/dsintro.html`

Grouping, Aggregation, and Correlation

The `pandas` library provides us with syntactic sugar for grouping and aggregating a DataFrame just by applying the `groupby` method and selecting a column for grouping:

```
g = iris.groupby("name")
for name, group in g: print name
>>>setosa
>>>versicolor
>>>virginica
```

In the following screenshot, we can see the aggregated data using the `sum`, `max`, and `min` methods for the dataset grouped by `name`:

In [11]: `iris.groupby("name").sum()`

Out[11]:

name	SepalLength	SepalWidth	PetalLength	PetalWidth
setosa	250.3	171.4	73.1	12.3
versicolor	296.8	138.5	213.0	66.3
virginica	329.4	148.7	277.6	101.3

In [12]: `iris.groupby("name").max()`

Out[12]:

name	SepalLength	SepalWidth	PetalLength	PetalWidth
setosa	5.8	4.4	1.9	0.6
versicolor	7.0	3.4	5.1	1.8
virginica	7.9	3.8	6.9	2.5

In [13]: `iris.groupby("name").min()`

Out[13]:

name	SepalLength	SepalWidth	PetalLength	PetalWidth
setosa	4.3	2.3	1.0	0.1
versicolor	4.9	2.0	3.0	1.0
virginica	4.9	2.2	4.5	1.4

We may also call the `describe` method for grouped data, as shown in the following screenshot. In this case, we will get aggregate data for each group:

```
In [14]: iris.groupby("name").describe()
Out[14]:
```

name		SepalLength	SepalWidth	PetalLength	PetalWidth
setosa	count	50.000000	50.000000	50.000000	50.000000
	mean	5.006000	3.428000	1.462000	0.246000
	std	0.352490	0.379064	0.173664	0.105386
	min	4.300000	2.300000	1.000000	0.100000
	25%	4.800000	3.200000	1.400000	0.200000
	50%	5.000000	3.400000	1.500000	0.200000
	75%	5.200000	3.675000	1.575000	0.300000
	max	5.800000	4.400000	1.900000	0.600000
versicolor	count	50.000000	50.000000	50.000000	50.000000
	mean	5.936000	2.770000	4.260000	1.326000
	std	0.516171	0.313798	0.469911	0.197753
	min	4.900000	2.000000	3.000000	1.000000
	25%	5.600000	2.525000	4.000000	1.200000
	50%	5.900000	2.800000	4.350000	1.300000
	75%	6.300000	3.000000	4.600000	1.500000
	max	7.000000	3.400000	5.100000	1.800000
virginica	count	50.000000	50.000000	50.000000	50.000000
	mean	6.588000	2.974000	5.552000	2.026000
	std	0.635880	0.322497	0.551895	0.274650
	min	4.900000	2.200000	4.500000	1.400000
	25%	6.225000	2.800000	5.100000	1.800000
	50%	6.500000	3.000000	5.550000	2.000000
	75%	6.900000	3.175000	5.875000	2.300000
	max	7.900000	3.800000	6.900000	2.500000

We can also group by multiple attributes, as shown in the following code:

```
for name, group in iris.groupby(["name", "SepalLength"]):
    print name
    print group
```

The resulting groups will look like this:

```
('setosa', 4.3)
        name  SepalLength  SepalWidth  PetalLength  PetalWidth
13   setosa          4.3           3          1.1         0.1
('setosa', 4.4)
        name  SepalLength  SepalWidth  PetalLength  PetalWidth
8    setosa          4.4         2.9          1.4         0.2
38   setosa          4.4         3.0          1.3         0.2
42   setosa          4.4         3.2          1.3         0.2
  . . .
```

> In the following link, we can find the Pandas groupby method
> documentation:
> http://pandas.pydata.org/pandas-docs/dev/groupby.html

The pandas library DataFrames provide us with a correlation function (corr) and implement three different correlation coefficient methods: pearson (default), kendall, and spearman, using the method parameter:

iris.corr(method='spearman')

In this case, we will get the correlation between two attributes (see In[15] in the following screenshot) and the correlation between all attributes (see In[16] in the following screenshot):

```
In [15]: iris["SepalLength"].corr(iris["PetalLength"])

Out[15]: 0.87175377588658287

In [16]: iris.corr()

Out[16]:
```

	SepalLength	SepalWidth	PetalLength	PetalWidth
SepalLength	1.000000	-0.117570	0.871754	0.817941
SepalWidth	-0.117570	1.000000	-0.428440	-0.366126
PetalLength	0.871754	-0.428440	1.000000	0.962865
PetalWidth	0.817941	-0.366126	0.962865	1.000000

Sharing your Notebook

One of the most amazing features of Wakari is that we can share our notebooks with other Wakari users and they can import them into their accounts. This feature makes Wakari an excellent choice for teaching a workshop or for a presentation.

The data

When our IPython Notebook is ready, we can share it with other Wakari users just by clicking on the **Share** button next to the name of our Notebook in the resources tab.

In the following screenshot, we can see the **Sharing** window, where we may change the **NAME** and add a **DESCRIPTION** to our notebook. For paid accounts, we can also include a password to keep our notebook private:

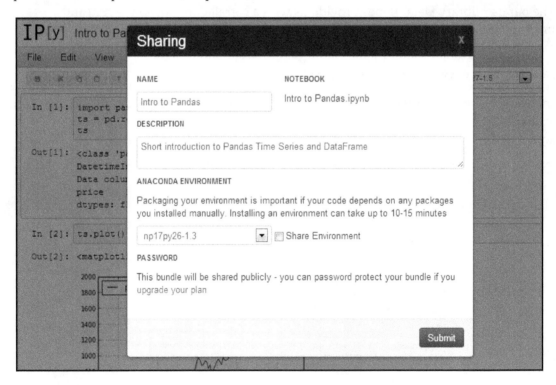

Once we are ready, we will click on the **Submit** button. We will see in the **Sharing Status** window that the process is complete, and we can click **Link to the bundle** to see our notebook being shared, as shown in the following screenshot:

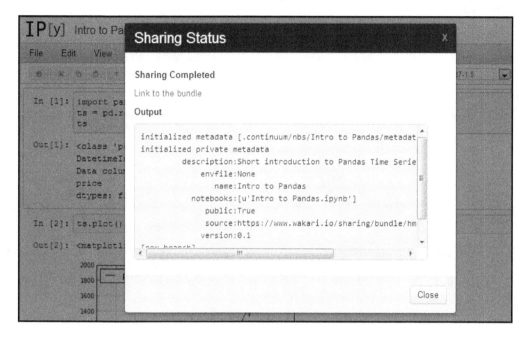

After clicking on **Link to the bundle,** we will see our IPython notebook **Intro to Pandas** as read-only. If we click on the **Run/Edit this Notebook** button, we will create a copy of the notebook in our Wakari environment , which we can upload freely, as shown in the following screenshot:

In the following screenshot, we can see our **Shared Bundles** by clicking on our **account name/Settings/Sharing;** there we can get the links or delete our shared notebook:

Wakari also provides us with a gallery in which we can find good tutorials as notebooks that we can copy and modify. We can find the gallery in the following link:

```
https://www.wakari.io/gallery
```

All the code and notebooks in this chapter may be found in the author's GitHub repository in the following link:
```
https://github.com/hmcuesta/PDA_Book/tree/master/Chapter14
```

Summary

In this chapter, we have explored an interesting tool for online data analysis with Python. Wakari provides us with scientific environments ready to use; this is a great tool for teaching and sharing code. In this chapter, we provided a short introduction to image processing and the `pandas` library. In `pandas`, we learned how to work with Time Series and with multivariate datasets. Finally, we learned how to share our IPython notebooks with other Wakari users and how IPython has evolved into the Jupyter project.

Wakari is highly recommended for all the Python community, because it provides a robust Anaconda environment out of the box and support for all the major Python libraries.

15
Understanding Data Processing using Apache Spark

In this chapter, we will present the main features of data processing architecture and the Cloudera platform distribution. Then, we will explore how to use a distributed filesystem and how to managing files from terminal and using a web interface. Finally, we will describe the use of **Apache Spark**, which is an open source, big data processing framework built with the goal of being fast and easy to use. Apache Spark provides us with a unified framework to manage big data processing requirements, such as data streaming, machine learning, and analytics.

In this chapter, we will cover these topics:

- Understanding data processing
- Platform for data processing
- An introduction to the distributed file system
- An introduction to Apache Spark
- Understanding data processing

Since the first edition of this book in 2013, there has been big changes in the data-driven scene. With the emerge of buzzwords such as big data, data science, and deep learning, the tools and methods have evolved. One of the main tools that has gained a great community and professional use is Apache Spark due to its velocity, adaptability with other cluster technologies such as **YARN** or **Apache Mesos**, and support for a lot of major languages such as Java, Scala, Python, and R.

The data architecture that we will explore in this chapter contains three elements:

- **Cluster Management**: This in charge of managing your cluster and scheduling jobs on your nodes.
- **Distributed storage**: Data processing involves a large volume of data. This data requires reliable and scalable data storage organized through a cluster. A distributed file system is used to simplify replication through a number of physical disks.
- **Process and analysis**: In this layer, we will implement data processing either in a batch (the data is processed after all input has been received) or streaming (the data is processed in real time).

Take a look at the following diagram:

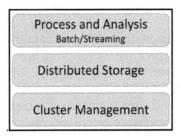

Platform for data processing

A data processing architecture requires a lot of technologies and tools for jobs scheduling until data streaming. A few years ago, installing all the tools required for data processing. Now we may implement the entire environment in just one step. Data companies such as **Cloudera**, **Hortonworks**, and **MapR** provide us with a complete setup of data environment in a Virtual Machine for a single-node cluster and a Docker container (automates the deployment of Linux applications inside software containers) for multi-node cluster.

Many data analytics applications need to process large datasets in batch post-processing or live streaming. For a data scientist, the time to set up a complete environment is a priority, and installing a ready setup platform is the best way to get hands on into the action. One of the main advantages here is that if you are working with either a single-node cluster or a multi-node cluster, the programming of Apache Spark projects are independent to the infrastructure.

The Cloudera platform

In this chapter, we chose the Cloudera distribution of big data platform due to its flexibility and support for Apache Spark. In the following distributed diagram, we can see the complete landscape of the data environment.

The platform integrates a lot of tools from the data integration of structured and unstructured data: storage in the form of file systems, NoSql, and relational databases. The platform support data processing in batch, streaming and querying. Finally, the platform unified cluster resource management, security, and scheduling. The following distributed diagram shows all the tools included in the Cloudera platform environment:

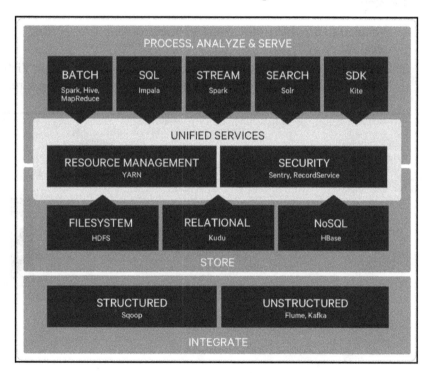

For a complete reference of QuickStart VM Software Versions and Documentation go to `http://bit.ly/2bGQb7Z`.

Installing Cloudera VM

In this case, we select a single-node cluster virtual machine of Cloudera distribution deployed on **Oracle VirtualBox**. The VM requires at least 8 GB of RAM in the host hardware and a minimum of 4 GB of RAM for the Virtual Machine of the **CDH5** version.

> We can find more information about `Virtualbox` installation from its website at `https://www.virtualbox.org/wiki/Downloads`.

Following are the steps:

1. We need to download the latest Virtualbox for our operating system and install it.
2. Once we have Virtualbox working, we need to create a new Virtual Machine with a Linux (64-bit), as is shown in the following screenshot:

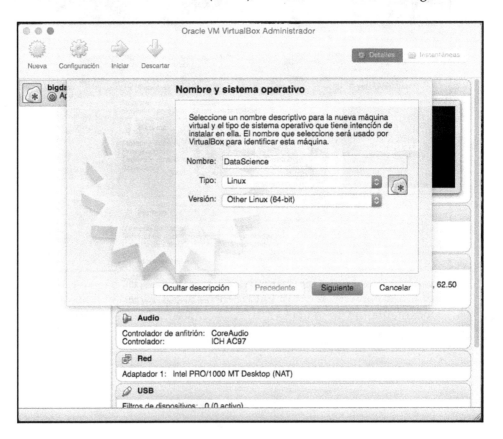

3. The steps to get the Cloudera VM up and running on VirtualBox are here:

> Press *Ctrl + I* and select the virtual image from Cloudera Push **Next** and push **import**, after a little time your image should be imported. Finally, select your virtual machine and press **Run**.

We can find the Cloudera VM distribution at the following link:
`http://www.cloudera.com/downloads/quickstart_vms/5-8.html`

4. Once we have finished the installation and running of our Cloudera VM, we will see the following screenshot:

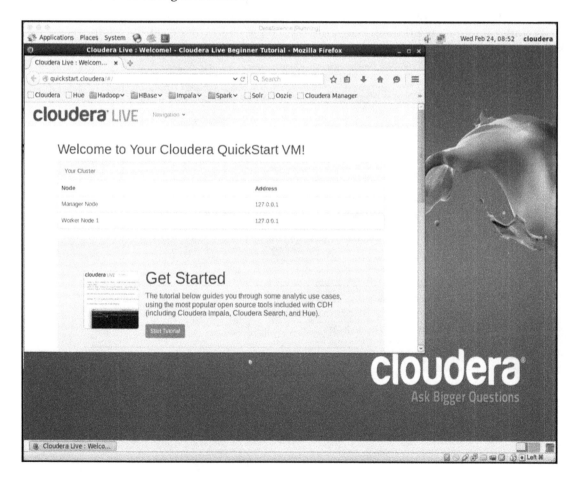

An introduction to the distributed file system

A distributed file system is practically the same as any file system due to its basic actions such as storing, reading, deleting files, and assigning security levels are support. The main difference is focused on the number of servers that can be used at same time without dealing with complexity of synchronization. In this case, we can store large files in different server nodes without caring about redundancy or parallel operations.

There are a lot of frameworks for distributed file systems, such as **Red Hat Cluster FS**, **Ceph File system**, **Hadoop Distributed File System (HDFS)**, and **Tachyon File System**.

In this chapter, we will use **HDFS**, which is an open source implementation of Google File System, built to handle large files into a cluster of commodity hardware. The HDFS cluster implements a **NameNode** that manages operations through the file system, and a series of **DataNodes** that manage the storage of the files in the cluster nodes individually, as is shown in the following diagram:

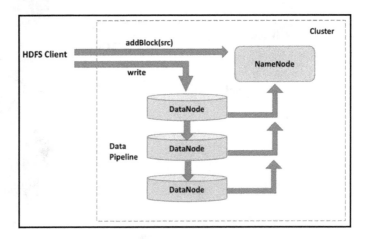

We can find more information about HDFS from:
http://www.aosabook.org/en/hdfs.html

First steps with Hadoop Distributed File System – HDFS

The most common way to use HDFS is through terminal commands. In general, we will manage texts or images files over 100 MB of size. We will interact with HDFS as a single disk; all the complexity is hidden by the tool:

1. Open a terminal and create a directory using this command:

```
hadoop fs -mkdir <paths>
>>> hadoop fs -mkdir /user/data
```

2. Copy a text file to the HDFS with this command:

```
hadoop fs -put <local-src> ... <HDFS_dest_path>
>>> hadoop fs -put texts.txt /user/data/folder/texts.txt
```

3. List the files in the new directory with this command:

```
hadoop  fs  -ls  <args>
>>> hadoop fs -ls /user/data/folder
```

4. Make a copy of the texts.txt file from HDFS to the local file system with the following command hadoop fs -get <hdfs_src> <localdst>:

```
>>> hadoop fs -get /user/data/folder/texts.txt /documents/
```

 We can find more information about the HDFS commands from the documentation in the following link:
http://bit.ly/2bQq6kF

File management with HUE – web interface

Hue is an open source web interface used to easily manage and analyze data with the Apache Hadoop ecosystem. We can handle your files in HDFS from the web interface, we just need to open the web browser on the Cloudera VM and enter `http://localhost:8888/` to open Hue, as shown in the following screenshot, and we will log in using `cloudera` as the username and password:

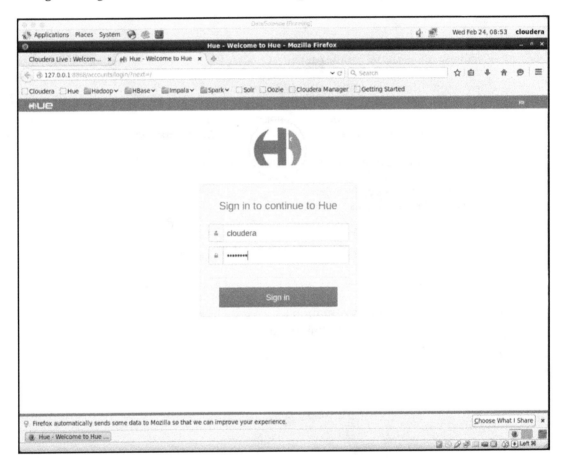

Once in the Hue interface, we can go to the **File Browser** option and upload and download files directly to the HDFS; this simplifies the managing and exploration of files and folders, as shown in the following screenshot:

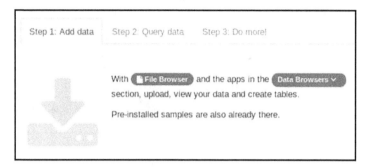

For reference documentation of the HUE web interface, go to the following link:
http://gethue.com/

An introduction to Apache Spark

Apache Spark is an open source cluster computer system with implicit data parallelism and fault tolerance. Spark was originally created at AMPlab from UC Berkeley; the main goal of Spark is to be fast to run and read and to apply in-memory processing. Spark allows you to manipulate distributed datasets, such as local collections. In this section, we will present the basic operations with Spark programming model and its ecosystem.

We can find more information about pandas from its website:
http://spark.apache.org/

The Spark ecosystem

Spark comes with a lot of high-level libraries for SQL querying, machine learning, graph processing, and streaming data. These libraries provide all inclusive environment ready to use. The following figure shows the complete Spark ecosystem:

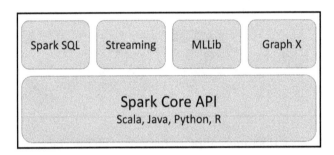

Take a look at the following:

- **Spark Core API**:

    The characteristics of Spark Core API are as follows:

    - It is the execution engine that allows all the other functionalities built on top
    - It provides an in-memory processing speeding up the execution of the jobs
    - It is built on the top of Scala programming language; however, it provides support for many other languages including Java, Python, and R

- **Spark SQL**:

    The characteristics of Spark SQL are as follows:

    - It is a programming abstraction to make interactive SQL-like queries for data exploration and reporting
    - It implements a data structure called **DataFrame** that works as a distributed SQL query engine
    - It also works as a data source for the other libraries of the Spark ecosystem, such as Spark ML for machine learning algorithms

- **Spark Streaming**:

  The characteristics of Spark Streaming are as follows:

  - It can consume streaming data; it is perfect when your data source is generating data in real time
  - It allows you to process data from different sources of living data streams, such as Twitter, Kafka, or Flume; and the process is fault tolerant

- **MLLib**:

  The characteristics of MLLib are as follows:

  - It's a scalable machine learning library
  - It includes algorithm implementation for most machine learning tasks, such as classification, regression, recommendation, optimization, and clustering

- **GraphX**:

  The characteristic of GraphX is as follows:

  - It is a graph computation engine that allows you to process graph-structured data; and it includes some of the most common algorithms for graphs such as centrality, distance.

> In the following link, we can find pandas Time Series Documentation:
> `http://pandas.pydata.org/pandas-docs/dev/timeseries.html`

The Spark programming model

We can program on Spark directly from the terminal using the **Spark shell** that returns an instant result as we enter the code. We can access the Scala-based Spark shell just by running `./bin/spark-shell` from the Spark base directory. In this chapter, we will use Python as the main language for Spark programming, so we will use the **PySpark** shell. So, we need to execute `./bin/pyspark` from the terminal in the base directory of Spark.

We are going to get a prompt ready for instructions, as shown in the following screenshot:

```
● ● ●                    spark-1.4.0 — java — 80×24
                              java                                      +
16/08/24 12:21:02 INFO Server: jetty-8.y.z-SNAPSHOT
16/08/24 12:21:02 INFO AbstractConnector: Started SelectChannelConnector@0.0.0.0
:4040
16/08/24 12:21:02 INFO Utils: Successfully started service 'SparkUI' on port 404
0.
16/08/24 12:21:02 INFO SparkUI: Started SparkUI at http://192.168.50.181:4040
16/08/24 12:21:02 INFO Executor: Starting executor ID driver on host localhost
16/08/24 12:21:02 INFO Utils: Successfully started service 'org.apache.spark.net
work.netty.NettyBlockTransferService' on port 51032.
16/08/24 12:21:02 INFO NettyBlockTransferService: Server created on 51032
16/08/24 12:21:02 INFO BlockManagerMaster: Trying to register BlockManager
16/08/24 12:21:02 INFO BlockManagerMasterEndpoint: Registering block manager loc
alhost:51032 with 265.1 MB RAM, BlockManagerId(driver, localhost, 51032)
16/08/24 12:21:02 INFO BlockManagerMaster: Registered BlockManager
Welcome to
      ____              __
     / __/__  ___ _____/ /__
    _\ \/ _ \/ _ `/ __/  '_/
   /__ / .__/\_,_/_/ /_/\_\   version 1.4.0
      /_/

Using Python version 2.7.12 (default, Jul  2 2016 17:43:17)
SparkContext available as sc, SQLContext available as sqlContext.
>>> ▊
```

The first step when we want to program in Spark is to create `SparkContext`, which is sets up with a `SparkConf` object that contains the cluster configuration settings, as if we were on a multi-core or single-core cluster. This can be initialized from the constructor of the `SparkConf` class, as is shown in the following code:

```
>>> from pyspark import SparkContext
>>> sc = SparkContext("local","Fitra")
```

The Spark main data structure is **Resilient Distributed Dataset** (**RDD**), which is a collection of objects that are distributed or partitioned across the cluster nodes. An RDD object is fault-tolerant. It has the ability to reconstruct itself if any of the nodes is down due to loss of communication or a hardware failure. We can create a new RDD object from an input file, as shown in the following code:

```
>>> data = sc.textFile("texts.txt")
```

The RDD object include a series of **actions** and **transformations** to process the data. The **actions** return a value, for example, `count()` that returns the number of records in the RDD or `first()` that returns the first record in the dataset:

```
>>> data.count()
10000
>>> data.first()
u'# First record of the text file'
```

The **transformations** return a new RDD with the result of the process; for example, if we want to filter the text in the RDD to find the records with the word `"second"`, we perform this:

```
>>> newRdd = data.filter(lambda line: "second" in line)
```

> In the following link, we can find Spark actions and transformations documentation:
> `http://bit.ly/1I5Jm9g`

When a program is running on Spark, the process runs a **SparkUI**, which is a web-based monitor of resources such as jobs, storage, and executors (nodes on execution). We can access this monitor with the `http://192.168.50.181:4040` URL. This is accessible while the process is running. For example, we can run an example of correlation calculus included with Spark by executing the example, and we can see the SparkUI of the process in the following screenshot:

```
>>>./bin/run-example org.apache.spark.examples.mllib.Correlations
```

← C ① 192.168.50.181:4040/executors/													

Spark 1.4.0   Jobs   Stages   Storage   Environment   Executors   Correlations

Executors (1)

Memory: 199.7 KB Used (265.1 MB Total)
Disk: 0.0 B Used

Executor ID	Address	RDD Blocks	Memory Used	Disk Used	Active Tasks	Failed Tasks	Complete Tasks	Total Tasks	Task Time	Input	Shuffle Read	Shuffle Write	Thread Dump
driver	localhost:50880	4	199.7 KB / 265.1 MB	0.0 B	1	0	5	6	2.5 s	266.1 KB	0.0 B	0.0 B	Thread Dump

One thing to mention is that all the transformations in Spark are **lazy evaluation** methods. This means that the transformations are only computed when an action needs a returned value.

An introductory working example of Apache Startup

In this section, we will explore one of the most classical examples for distributed programming, the *word count*, where we will understand how many times each word appears. In this example, we are going to implement a map and a reduce method that we have already seen in Chapter 13, *Working with MapReduce*. In this case, we will use a Spark implementation in Python, using the map action and the reduceByKey transformation:

Here are the steps for implementation:

1. Import SparkContext from the pyspark library.
2. From pyspark, import SparkContext.
3. Create a SparkContext object using a single-node configuration (local).
4. Assign a name to the job (WCount):

```
sc = SparkContext("local","WCount")
```

5. Load a text file from the HDFS into a RDD objet named textFile:

```
url = "hdfs://localhost/user/cloudera/words.txt"
textFile = sc.textFile(url)
```

6. Create a vector with the word and the number of times it appears into the text file.
7. Create a vector with all the words individually with the transformation called flatMap(lambda line: line.split(" ")).
8. Then, with the .map(lambda word: (word, 1)) transformation, we will define how many times a word appears, adding the number 1 to the vector of each word.
9. Finally, with the .reduceByKey(lambda a, b: a + b) action, we will add all the times that a word appears to get the result:

```
counts = textFile.flatMap(lambda line: line.split(" ")) \
            .map(lambda word: (word, 1)) \
            .reduceByKey(lambda a, b: a + b)
```

10. Finally, we need to save the result into a folder with the saveAsTextFile function:

```
counts.saveAsTextFile("hdfs://localhost/user/cloudera/out")
```

To execute the code, we need to go to the terminal and run the `./bin/spark-submit <name of the file>` command, and we can see the result in HDFS in the folder out, where we are going to find to the `_SUCCESS` files that show the status of the execution. We will also find a text file called `part-00000` where we can see the result, as shown in the following screenshot:

```
>>> ./bin/spark-submit TestSpark.py
```

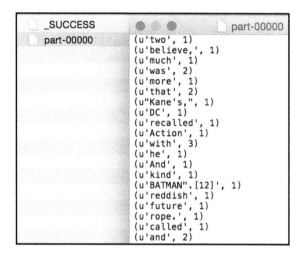

All the code and notebooks of this chapter can be found in the author's GitHub repository at the following link:
`https://github.com/hmcuesta/PDA_Book/Chapter15`

Summary

In this chapter, we explored briefly how to understand a data processing architecture. First, we explored how to interact with a distributed file system. Then, we provided installation instructions for a Cloudera VM and how to get started in a data environment. Finally, we described the main features of Apache Spark and ran a practical example of how to implement a word count algorithm.

Apache Spark is highly recommended for all the data community, because it provides a robust and fast data processing tool. It also provides a library for Sql-like querying, graph processing, and machine learning algorithms.

Index

www.ingramcontent.com/pod-product-compliance
Lightning Source LLC
Chambersburg PA
CBHW062059050326
40690CB00016B/3150